Dictionary of Farm Animal Behavior

SECOND EDITION

T0255831

Dictionary of Farm Animal Behavior

SECOND EDITION

J.F. Hurnik
A.B. Webster
P.B. Siegel

IOWA STATE UNIVERSITY PRESS / AMES

J. Frank Hurnik, PhD, is Professor in the Department of Animal and Poultry Science, University of Guelph, Ontario, Canada.

A. Bruce Webster, PhD, is Professor in the Department of Poultry Science, The University of Georgia, Athens.

Paul B. Siegel, PhD, is Professor in the Department of Animal and Poultry Science, Virginia Polytechnic Institute and State University, Blacksburg.

First edition © 1985 University of Guelph, Ontario, Canada N1G 2W1

Second edition © 1995 Iowa State University Press, Ames, Iowa 50014
All rights reserved

Authorization to photocopy items for internal or personal use, or the internal or personal use of specific clients, is granted by Iowa State University Press, provided that the base fee of $.10 per copy is paid directly to the Copyright Clearance Center, 27 Congress Street, Salem, MA 01970. For those organizations that have been granted a photocopy license by CCC, a separate system of payments has been arranged. The fee code for users of the Transactional Reporting Service is 0-8138-2464-8/95 $.10.

Second edition, 1995

Library of Congress Cataloging-in-Publication Data

Hurnik, J. F. (Jaromir Frank)
    Dictionary of farm animal behavior / J. F. Hurnik, A. B. Webster, P. B. Siegel.—2nd ed.
        p. cm.
    ISBN: 978-0-8138-2464-2
    1. Livestock—Behavior—Dictionaries. I. Webster, A. B. (Arthur Bruce). II. Siegel, P. B. (Paul Benjamin).
III. Title.
SF756.7.H87   1995
599'.051—dc20   95-16681

Contents

Preface, vii

Acknowledgments, ix

Alphabetical listing, 3

Preface

IN MODERN AGRICULTURE, there is an increasing awareness that animal behavior is relevant to animal production. Educational institutions have introduced courses in behavior to promote understanding of the relationship between the behavior, productivity, and well-being of farm animals. Additionally, an increased amount of research is being initiated to elucidate behavioral principles that ultimately may be incorporated into the design of production systems to improve productivity and minimize stress of farm animals.

The authors believe a dictionary of farm animal behavior is necessary for two reasons. First, as in any area of rapid scientific growth, the terminology of farm animal behavior has proliferated, and its usage in many cases is not standardized. Secondly, students in programs related to animal agriculture are generally unfamiliar with behavioral terminology, and there is need to provide materials that will facilitate their learning. This edition is an extended version of previously published *Dictionary of Farm Animal Behavior* (1985). New entries have been included to cover the basic terminology of animal welfare and agroethics.

During the development of the dictionary, considerable time and effort went into composing straightforward, unambiguous definitions for words that often are used imprecisely. Many terms representing behavioral principles that apply to, but also go beyond, the realm of animal agriculture have been included, and numerous colloquial expressions have been added with reference, when possible, to more rigorous technical terms. It has been our goal to provide a reference book for behavioral terminology and ethological concepts that is reasonably comprehensive yet comfortably sized and relatively inexpensive. Students of behavior, animal scientists, veterinarians, and others who are interested in or who work or intend to work with animals as a career or vocation should find the dictionary to be a useful resource.

As in any human endeavor, we recognize the possibility that, despite our best efforts, some important terms may have been omitted or some definitions may be incomplete or unclear. The authors would appreciate these being brought to their attention so that appropriate revisions can be made to any future edition.

Acknowledgments

THE AUTHORS GRATEFULLY ACKNOWLEDGE Drs. Ed Bailey, Joanna Boehnert, Hugh Lehman, and Russ Willoughby for their constructive criticism and suggestions regarding terms relating to their respective fields. For effective editing and typing assistance we would like to express our appreciation to Sandra Brown, Mary Cocivera, Ian Easterbrook, Bev Gurr, Gill Joseph, Mary-Ellen Pyear, Gwen Scott, Peter Taylor, and Sheila Ward. We owe special recognition to the editors and staff of Iowa State University Press, especially Gretchen Van Houten. The cover and inside page illustrations were done by Leslie Richards.

Dictionary of Farm Animal Behavior

SECOND EDITION

A

ABANDONMENT CALL: *See* Separation Call.

ABDUCENS NERVE: The sixth cranial nerve which innervates the lateral rectus and retractor bulbi muscles. Lesions of this nerve cause the eyeball to deviate medially.

ABDUCTION: Movement away from the median plane.

ABERRANT BEHAVIOR: *See* Abnormal Behavior.

ABERRATION: Deviation from that which is typical, regular, common, or standard.

ABIENT: Avoiding or moving away from the source of stimulation. Antonym: Adient.

ABILITY: Competence of an organism to perform a given task. *Also see* specific ability—MOTHERING.

ABLACTATION: Cessation of milk secretion.

ABNORMAL: Deviating from a norm. The term has many applications, the most objective being a statistical interpretation focusing on qualitative or quantitative deviations from expected frequencies, means, or intervals.

ABNORMAL BEHAVIOR: Behavior that deviates in form, frequency, or sequence from a defined, comparable standard. Such a standard may be a behavioral inventory typical for a given genotype, age group, sex, nutritional level, housing condition, or management system, etc.

ABORAL: The region of the body considered opposite to the mouth, or in a direction of movement away from the mouth (e.g., along the digestive tract).

ABORTION: Premature expulsion of offspring from the uterus, either causing their death or because of it.

ABSCESS: A localized occurrence of pus in a cavity of disintegrated tissues.

ABSOLUTE SCALE: A scale that starts at absolute zero and extends in regular intervals over the whole range of a given variable.

ABSOLUTE SENSITIVITY: Ability to respond to stimuli of any intensity.

ABSTRACTION: A concept containing the essential elements of a larger or more complex entity; also, the development of such a concept.

ACCEPTANCE, SOCIAL: *See* Social Acceptance

ACCESSORY NERVE: The eleventh cranial nerve which innervates the palate, larynx, and muscles of the neck and thorax.

ACCESSORY ORGAN: An additional organ(s) which facilitates processes carried out by specialized systems in the body (e.g., the muscles of the eye are accessory organs in vision).

ACCIDENTAL REINFORCEMENT: *See* Incidental Reinforcement.

3

ACCLIMATION: The reversible ontogenic process of becoming accustomed to an artificially controlled change in a single or limited number of environmental factors.

ACCLIMATIZATION: The reversible ontogenic process of becoming accustomed to changes in a wide range of environmental factors independent of direct human control.

ACCOMMODATION: The relatively rapid adjustment of an organism, organ or part of an organ to an existing situation. Accommodation takes place on a shorter time scale than acclimation or acclimatization.

ACHROMATIC: Absence of color (chroma) encompassed by the chromatic spectrum.

ACHROMATOPSIA: Inability to discriminate between colors (color blindness).

ACOUSTIC: Referring to hearing and sounds.

ACOUSTIC NERVE: *See* Auditory Nerve.

ACQUISITION: Behavior modifications which enlarge the behavioral inventory of an organism.

ACROMANIA: A mental disorder marked by great motor activity.

ACT, CONSUMMATORY: *See* Consummatory Act.

ACTION: *See* specific action—BEHAVIORAL; SYMBOLIC.

ACTION CHAIN: Any sequence of actions which is executed in a nonrandom order.

ACTION CURRENT: Electric current associated with the passage of a wave of excitation along a nerve or muscle fiber.

ACTION PATTERN: A series of behavioral actions having consistent form and sequence when performed by an organism. If the action pattern is typical of a given species or breed, it may be designated as a modal action pattern or a fixed action pattern.

ACTION PATTERN, FIXED: *See* Fixed Action Pattern.

ACTION PATTERN, MODAL: *See* Modal Action Pattern.

ACTION POTENTIAL: An abrupt, short-term (a few thousandth's of a second), pulse-like change in ion polarity which moves along the length of a nerve, thus giving rise to a nerve signal. The action potential occurs in two separate stages called depolarization and repolarization.

ACTION-SPECIFIC ENERGY: Energy that provides the motivational basis for the expression of a fixed action pattern. Each behavioral pattern would be driven by its own source of action-specific energy. The concept of action-specific energy largely has fallen out of usage.

ACTIVATION: *See* Arousal.

ACTIVITY: A behavioral state, or arbitrarily defined group of such states. The term generally is used with an adjective to denote some qualitative or quantitative aspect of the behavioral state(s) (e.g., feeding activity, high activity).

Also see specific activity—DOMINANT; INTERIM; PIECEMEAL; SPONTANEOUS; SUBDOMINANT.

ACTIVITY RANGE: The area covered by an animal in the course of its regular daily activities.

ACUITY: Sharpness, particularly referring to sensory perception of low intensity stimuli.

ACUTE DISEASE: Any disease that appears rapidly regardless of whether it leads to recovery or death. *Compare:* Chronic Disease.

ADAPTATION: The phylogenetic process of a species becoming better adjusted to its environment through genetic selection. Adaptation also is used to refer to short-term changes in sensitivity to stimulation. *Compare*: Sensory Adaptation.

ADAPTEDNESS: *See* Adaptive Value.

ADAPTIVE BEHAVIOR: Behavior that has a high adaptive value.

ADAPTIVE COMPLEX: The complete array of behavioral, physiological, morphological, and psychological traits involved in the process of adaptation.

ADAPTIVE SIGNIFICANCE: *See* Adaptive Value.

ADAPTIVE VALUE: Assessment of a trait (behavioral, physiological, morphological, etc.) with regard to its contribution to genetic fitness. *Compare*: Survival Value.

ADDUCTION: Movement toward the median plane.

ADEQUATE STIMULUS: In experimental situations a term used occasionally to refer to a stimulus that activates sensory receptors considered to be appropriate for such a stimulus. *Compare*: Inadequate Stimulus.

ADIENT: Moving toward the source of stimulation. Antonym: Abient.

ADIPOSIS: Obesity; excessive accumulation of fat in the body.

ADIPSIA: Complete cessation of drinking. It can be experimentally induced by lesions in the lateral hypothalamus. Antonym: Polydipsia.

ADJUNCTIVE BEHAVIOR: Behavior induced by an intermittent reinforcement schedule characteristically displayed during the time interval when reinforcement is highly improbable.

AD LIBITUM (AD LIB.): At pleasure (commonly used to indicate unrestricted access to feed or water).

ADOPTION: Acceptance of alien offspring, usually as infants, by an adult that provides them with parental care.

ADRENALIN: *See* Epinephrine.

ADRENOCORTICOTROPIC HORMONE (ACTH): A hormone released from the anterior lobe of the pituitary gland in response to the corticotropic-releasing factor. ACTH stimulates the adrenal cortex to produce cortical hormones (corticosterone, cortisone, etc.) and to decrease the level of ascorbic acid in the adrenal cortex.

ADULT SUCKLING: Suckling of a lactating female by an adult conspecific.

Habitual adult suckling in farm animals is classified as a social vice.

ADVENTITIOUS REWARD: A reward that is unplanned by the experimenter, or not under the experimenter's control, but that influences the behavior of the test organism.

ADVERTISEMENT: Any form of communication displayed to attract conspecifics (typically potential sexual partners) or to distract, warn, or threaten potential rivals or adversaries.

AEROPHAGIA: The swallowing of air. Excessive aerophagia, commonly occurring during habitual cribbing in horses, can lead to serious digestive problems and is considered a dangerous vice.

AFFECTION: Behavior that indicates positive feelings toward other organisms of the same or different species.

AFFECTIVITY: A tendency to react emotionally.

AFFERENT: Carrying toward, as in the transmission of neural messages toward the central nervous system.

AFFERENT INHIBITION: Temporary cessation of response to a given stimulus as a result of repetition of the stimulus. If the stimulus is slightly modified, or if it is repeated after a period of rest, the response reappears.

AFFERENT NERVES: Nerves that transmit action potentials from sensory receptors. Also called sensory nerves. *Compare*: Efferent Nerves.

AFTERBIRTH: The placenta and placental membranes expelled from the uterus following parturition.

AFTER-IMAGE: Perception of an image of a visual stimulus after the stimulus is withdrawn.

AFTERIMPRESSION: *See* Aftersensation.

AFTERSENSATION: A sensation outlasting the duration of the stimulation which induced it.

AGALACTIA: Cessation of milk secretion in postpartum females shortly after parturition.

AGENT, MORAL: *See* Moral Agent.

AGEUSIA: Undeveloped or seriously underdeveloped sense of taste.

AGGREGATION: Clustering of organisms independently attracted to some environmental resource (e.g., feed, water, temperature, etc.). Aggregation is distinguished from grouping induced by social factors. *Compare*: Flocking, Herding.

AGGRESSION: Any purposive action of an organism toward another organism with the actual or potential result of harming, limiting, or depriving it. In special cases, a nonliving object or the instigator itself may be the focus of the aggressive action. *Also see* specific aggressions—DEFENSIVE; DOMINANCE; OFFENSIVE; REINFORCED; SEXUAL; TERRITORIAL.

AGGRESSIVE BEHAVIOR: *See* Aggression.

AGGRESSIVITY VALVE: A hypothetical mechanism involved in the harmless discharge of aggressive motivation.

AGITATED DEPRESSION: Depression characterized by hyperactivity and restlessness.

AGITATION: Relatively strong emotionality, nervousness, or restlessness.

AGLUTITION: Inability to swallow.

AGNOSIA: Inability to perceive; generally used in reference to specific sensory modalities, such as auditory, gustatory, olfactory, tactile, or visual agnosia.

AGONIST: A muscle that initiates action and is opposed by another muscle. Antonym: Antagonist.

AGONISTIC: Refers to any activity performed in the context of an aggressive interaction. It encompasses the actions of both the instigator and the victim.

AGONISTIC BEHAVIOR: Any behavior indicative of social conflict such as threat, attack, and fight; or escape, avoidance, appeasement, and subordination.

AGONY: A state of extreme suffering and distress often associated with moribund behavior.

AGRICULTURAL ANIMAL: A domesticated animal used in agriculture.

AGRICULTURE: The practice of establishing and nurturing populations of domesticated plants or animals to produce products (primarily foodstuffs) for human purposes.

AGROETHICS: Branch of bioethics focusing on ethical issues in agriculture and agricultural research, such as sustainability of food production, affordability of food, socioeconomic impact of new technologies, and ethical treatment of farm animals.

AHEMERAL CYCLE: A light-dark cycle other than 24 hours in length.

AHIMSA: A religious doctrine stating that it is wrong to kill any living thing. This doctrine is based on the belief that souls transmigrate and may inhabit any living being.

AIDS (horse): Signals produced by a trainer or rider to communicate with a horse. Some people differentiate between natural aids (hands, legs, body, and voice) and artificial aids (whips, spurs, and martindale).

AIR PECKING (poultry): Pecking movements toward no obvious target.

AKINESIA: Absence or severely reduced capability of movement.

ALARM CALL: A vocal alarm signal.

ALARM REACTION: The first stage of the general adaptation syndrome. *Also see* General Adaptation Syndrome.

ALARM RESPONSE: Any behavioral response indicative of fear or awareness of danger.

ALARM SIGNAL: A signal emitted by an organism to alert other individuals in the vicinity of the presence of danger. The signal has a high arousal potential for conspecifics, and often for other animals as well.

ALERT: The state characterized by high attentiveness and responsiveness to stimuli. *Compare*: Consciousness, Awareness.

ALG(O)-: A prefix referring to pain.

ALGESIA: Abnormally high sensitivity to pain. Antonym: Analgesia.

ALGESIOMETER: An instrument used to measure sensitivity to pain.

ALIMENTARY: Referring to organs of digestion.

ALLELE: One of two or more alternate forms of a gene occupying a given gene locus. Different alleles code for different manifestations of the trait that they control or influence.

ALLELOMIMETIC BEHAVIOR: Behavioral activities that have strong components of social facilitation, imitation, and group coordination. Synonym: Allomimetic Behavior.

ALLIANCE (horse): A form of cooperation between stallions in a multi-male bands of feral horses. One stallion may confront an approaching male from another band, while the other stays with his group. On the next such confrontation the roles of the two stallions might change.

ALLOCHTHONOUS BEHAVIOR: A term arising out of the concept of action-specific energy. Allochthonous behavior is that which is driven by energy that has sparked over from some other drive due to thwarting of the expression of behavior specific to the latter drive. In this conceptual system, displacement activities could be considered allochthonous behaviors. *Compare*: Autochthonous Behavior.

ALLOGROOMING: *See* Grooming.

ALLOKINESIS: Involuntary motion that occurs in simple reflexes induced by external stimuli.

ALLOLICKING: *See* Licking.

ALLOMIMETIC BEHAVIOR: *See* Allelomimetic Behavior.

ALLOMONES: Biochemical substances emitted by an individual of one species to communicate with an individual of another species. Allomones are typically transmitted by air.

ALLOMOTHER: A female that assumes the epimeletic role of a dam on behalf of young that are not her own offspring.

ALLOPARENT: An individual other than a natural parent that displays parental care.

ALLOPATRY: Geographical separation of populations. Allopatric populations of a species may undergo genetic differentiation if the period of separation extends over a large number of generations.

ALLORHYTHMIA: Irregular rhythmicity; used most frequently in reference to heart beat.

ALL-OR-NONE RESPONSE: A response that is either elicited or not and, if elicited, shows no grading.

ALLOTRIOPHAGIA: *See* Pica.

ALPHA ACTIVITY: *See* Brain Wave Activity.

ALPHA ANIMAL: The animal that ranks highest socially in its group (Animals in an established linear social hierarchy are often designated with letters of

A

the Greek alphabet according to their rank.) *Compare*: Omega Animal.

ALTRICIAL: Refers to species whose individuals are insufficiently developed at birth or hatching to see, move in a coordinated fashion, and fend for themselves. Neonatal individuals require considerable parental care. *Compare*: Precocial.

ALTRIGENDERISM: Relates to nonsexual activities that occur between individuals of opposite sex.

ALTRUISM: A phenomenon in which one organism does something to the benefit of another organism(s), usually at some cost to itself. In the context of human conduct, altruistic moral philosophy asserts that morality cannot be based exclusively on satisfaction of self-interest.

ALTRUISTIC BEHAVIOR: Behavior by which an organism manifests altruism.

AMATORY BEHAVIOR: A term occasionally used to refer to activities indicative of the development and maintenance of attachment between animals. Behavioral signs include allogrooming, muzzling, licking, and social play.

AMAUROSIS: Complete blindness.

AMBIENT TEMPERATURE: A term used loosely to refer to the air temperature of a given environment.

AMBIGUOUS: A statement or occurrence that may have two or more meanings. In a behavioral context this term is used in motivational assessment to refer to behavioral actions which have several different possible explanations or functions.

AMBIVALENT POSTURE: A form of compromise behavior in which an animal adopts a posture having separate elements suggestive of the influence of different motivations.

AMBLE (horse): A variation of the pace gait that is distinguished by a slight hesitation between the placing of the hooves of the front and rear legs on the same side of the horse. The amble is a slower gait than is the pace, but it is easier for the rider.

AMBLYOPIA: Reduced clarity of vision.

AMBULATION: Walking.

AMENSALISM: A form of interspecies relationship in which one species is affected adversely while the other remains unaffected.

AMNESIA: Partial or total loss of memory.

ANADIPSIA: Intense thirst.

ANALGESIA: Inability to experience pain. Antonym: Algesia.

ANALOGOUS BEHAVIOR: Behavior manifested by different species that is similar in function but does not originate from common ancestry. *Compare*: Homologous Behavior.

ANALOGY: A similarity between two events or things. Reasoning by analogy is to conclude that because two entities share one or more common charac-teristics they share another characteristic. For example, since a prerequisite for perception of a pain is the presence of specific, functionally active neu-rophysiological structures, all organisms which possess these functionally active structures are considered able to perceive pain.

ANALOGY, LAW OF: *See* Law of Analogy.

ANAL REFLEX: Spasmodic contraction of the anus in response to tactile stim-ulation.

ANALYSIS: *See* specific analysis—CAUSAL; COST-BENEFIT; FUNCTIONAL; MOTI-VATIONAL.

ANAPHIA: Lack of sense of touch.

ANDROGENS: A group of hormones that promote development of male sec-ondary sexual characteristics and male sexual behavior, and contribute to body growth of males. Androgens are produced mainly by interstitial cells (Leydig cells) of the testes, but are found also in the placenta, adrenal cor-tex, and ovaries.

ANECDOTAL EVIDENCE: Incidental evidence that does not provide suffi-cient grounds for generalization.

ANEMOTAXIS: Taxis related to direction of air flow.

ANESTHESIA: Loss of sensation usually produced by administration of a chemical substance(s). Anesthesia can be local, influencing only part of the organism, or general, influencing the organism as a whole. *Also see* specific anesthesia—CONDUCTION.

ANESTRUS: A relatively long period of sexual inactivity in females of repro-ductive age, generally occurring in response to seasonal change. *Compare*: Seasonal Breeding.

ANHYDROSIS: Failure of the system that controls sweating as a consequence of limited ability to acclimatize to a hot climate. Additional symptoms are dyspnea and dry skin having low elasticity. Anhydrosis occurs primarily in horses but also may occur in cattle. Synonym: Nonsweating Syndrome, Dry Coat.

ANIMAL: *See* specific animals—ALPHA; BOSS; DOMESTICATED; FARM; FERAL; FOOD; KASPAR-HAUSER; MARKER; OMEGA; PUREBRED; TIMID; TRACTABLE; UN-DOMESTICATED; WILD.

ANIMALISM: A theory that rejects any distinction between humans and ani-mals based on supernatural concepts.

ANIMAL RIGHTS: *See* Rights.

ANIMAL SCIENCE: The scientific study of animal life. In common use, ani-mal science refers to the range of applied scientific disciplines dealing with agricultural animals.

ANIMAL SOUL: An analogue of the human soul attributed to animals. Theories that assert the existence of animal souls claim that behavioral ac-tivities of animals cannot be fully explained by sole reference to physiological processes without reference to mental involvement indicative of the existence of a soul. This theory opposes the view that only humans possess a soul. *Compare*: Soul.

ANIMAL SUFFERING: *See* Suffering.

ANIMAL WELFARE MOVEMENT: Broad spectrum of sociopolitical activities concerned with living conditions of animals under human control or influence. The main objective of this movement is to prevent animal suffering and promote animal well-being.

ANIMAL WELFARE SCIENCE: Fields of scientific endeavor concerned with understanding what factors affect and what symptoms indicate animal suffering and animal well-being. Such fields include, but are not limited to, physiology, psychology, ethology, management, and health of animals.

ANIMAL WELL-BEING: *See* Well-being.

ANIMATION, SUSPENDED: *See* Suspended Animation.

ANIMISM: A term attributed to primitive, rudimentary forms of panpsychism.

ANNOUNCING (chicken): A vocalization emitted frequently after oviposition. Synonym: Cackling.

ANNOYER: An unpleasant stimulus.

ANOMALOUS BEHAVIOR: *See* Abnormal Behavior.

ANOREXIA: Chronic inappetance or reduced appetite for food.

ANOSMIA: Undeveloped sensitivity to olfactory stimuli.

ANOXIA: Lack of oxygen or reduction of oxygen levels in body tissues below normal physiological levels (e.g., altitude anoxia, myocardial anoxia).

ANTAGONISM: Mutual opposition. In the context of interspecies relationships, this term refers to situations in which two or more species affect each other adversely (e.g., by competing for the same type of food).

ANTAGONIST: A muscle that opposes the action of another muscle. Antonym: Agonist.

ANTE PARTUM: Before parturition, or more specifically, before onset of labor.

ANTERIOR: Situated in front of, or in the direction of, the head of the body.

ANTERIOR PRESENTATION: Fetal presentation in which the forelegs and nose jointly enter the birth canal. There are also two variations of anterior presentation: partially anterior presentation and cranial presentation.

ANTHRAX: An acute disease of cattle, sheep, goats, horses, pigs, and less frequently, poultry, caused by *Bacillus anthracis*. Behavioral symptoms include difficult respiration accompanied by high fever, indication of abdominal pain, unconsciousness, and eventually death. The symptoms appear in rapid succession and usually go undetected prior to death.

ANTHROPOMORPHISM: Attribution of human characteristics, abilities or priorities to nonhuman organisms.

ANTICIPATORY REACTION: A response to a stimulus before its actual occurrence.

ANTIDIURETIC HORMONE (ADH): A hormone secreted by the hypothalamus that inhibits diuresis, stimulates contraction of arterioles, and controls water loss from the kidney. Release of ADH is stimulated by increased osmotic pressure of blood, which is detected by the osmoreceptors of the hypothalamus. Synonym: Vasopressin.

ANTIPREDATOR BEHAVIOR: Any action having the purpose of reducing attacks by predators or diminishing their harm to an individual or group. Antipredator behavior includes cryptic behavior, vigilance, avoidance or escape, grouping, temporary group dispersion, defensive formation, selection of protective nest site, distraction of the predator from the nest site or offspring, threat display, discharge of noxious substances, and attack.

ANURIA: Absence of or complete retention of urine. *Compare*: Oliguria.

ANVIL: *See* Tympanic Ossicles.

APATHY: Listless and indifferent behavior; commonly a reaction to persistent and insurmountable frustration.

APERIODIC REINFORCEMENT: *See* Random Reinforcement.

APHAGIA: Refusal to eat.

APHRODISIA: Exaggerated sexual desire or unusually frequent display of sexual behavior.

APNEA: Lack of respiration. An apneic period occurs in newborns prior to initiation of regular pulmonary activity (neonatal apnea). Apnea also occurs as a consequence of neural malfunction (sleep apnea), or intense traumatic experience (traumatic apnea).

APOGEOTROPISM: Orienting response away from gravity. Antonym: Geotropism.

APOPATHETIC BEHAVIOR: Behavior influenced by the presence of conspecifics, but not directed toward them.

APOPLEXY, SPLENIC: *See* Anthrax.

APOSEMATIC: Conspicuous display of colored structures that increase the effect of threat and benefit self-defense capacity of animals.

APOSIA: Refusal to drink.

APOSITIA: Aversion to food.

APPARATUS, DOPPLER: *See* Doppler Apparatus.

APPARENT MOVEMENT: Subjective and illusionary visual perception of movement in the absence of real movement. Apparent movement can be generated by rapid succession of motionless stimuli that mimic the changes that occur in true movement.

APPEASEMENT: Conciliation directed toward an aggressor or potential aggressor.

APPEASEMENT SIGNAL: Any behavioral display indicative of conciliatory intent. Appeasement signals often are manifested toward threatening conspecifics when escape is either difficult or impossible.

APPERCEPTION: The process of conscious perception of events and full awareness, including self-awareness.

APPETITE: Desire for some commodity that can be neutralized by specific consummatory responses. The term generally is used in reference to recurrent desires related to physiological processes. *Also see* specific appetites— DIMINISHED; PERVERTED; REDUCED; RETURNED; VARIABLE.

APPETITIVE BEHAVIOR: Behavior manifested during the initial phase of an operant behavioral cycle indicative of desire to attain a certain goal (e.g., searching for food, pressing bar, broadcasting attraction signals, etc.).

APPETITIVE PHASE: The initial phase of an operant behavioral cycle. *Compare*: Behavioral Cycle.

APPLIED ETHOLOGY: The study of animal behavior conducted primarily for utilitarian purposes.

APPREHENSIVE BEHAVIOR: Any behavior indicative of an organism's anticipation of some adverse experience. The most common indicators include high level of alertness, intense sensory focusing on fear-causing stimulus(i), lowered "Flight or Fight" thresholds, defensive posture, and threat display.

APRAXIA: Inability of an organism to execute purposeful movements in order to cope effectively with its environment (e.g., after sustaining a brain lesion).

APROSEXIA: Inability to maintain focused attention.

APSELAPHESIA: Distortion of perception of touch.

APTITUDE: Capacity to perform a new, unlearned act.

ARC, NEURAL: *See* Neural Circuit.

ARCHED NECK (horse): Distinct flexion of the neck displayed during agonistic encounters between males and during courtship when a male is approaching a female.

AREA: *See* specific areas—EXERCISE; RESTING.

AREFLEXIA: Inability to execute a reflex act.

ARHIGOSIS: Inability to detect coldness.

AROUSAL: A dimension of neural activity against which behavioral and physiological events can be ranked. The term activation often is used interchangeably with arousal.

ARREST: Sudden cessation. This term is used frequently in reference to the activities or the function of specific organs (e.g., maturation arrest, cardiac arrest).

ARRHYTHMIA: *See* Allorhythmia.

ARTHRITIS: Inflammation of the joint, characterized by progressive locomotory difficulty and increasing time spent in a recumbent position with the affected joint flexed.

ARTIFICIAL BREEDING: Breeding which artificially replaces or by-passes one or more steps of natural breeding.

ARTIFICIAL INSEMINATION: Insemination mediated by humans whereby semen previously collected from a male is artificially transferred into the female reproductive tract.

ARTIFICIAL SELECTION: *See* Selection.

ASEXUALITY: Long lasting or permanent absence of sexual excitability and sexual activity.

ASPERGILLOSIS: A respiratory disease of birds and mammals caused by the fungus, *Aspergillus*. Behavioral symptoms include difficulty in breathing, sneezing, coughing, rattling noise in the throat, nasal discharge, and reduced appetite.

ASSOCIATION: In the context of learning, any cognitive connection between events. Association also refers to the involvement of organisms with each other in social relationships.

ASSOCIATION TEST: Any experimental technique to assess the reaction of a subject to a specific stimulus (e.g., a certain sound, color, etc.).

ASSOCIATIVE INHIBITION: Severe weakening of an established association due to a newly formed association. The term is used occasionally to refer to severe difficulty in establishing an additional association because of an existing association.

ASSOCIATIVE STRENGTH: Strength of association defined by the frequency with which a given stimulus evokes a particular conditioned response.

ASSORTATIVE MATING: A type of mating system in which choice of sexual partner is influenced by phenotypic resemblance. Assortative mating is called "positive" when individuals in mated pairs are of similar phenotype more frequently than random expectation, and "negative" when the reverse is true.

ASTHENIA: Lack of strength or unusual weakness of an organ or organism.

ASTROTAXIS: Locomotion related to the position of stars or other celestial bodies.

ASYMMETRIC GAIT: A gait in which the action of one or more legs is not synchronized with that of another leg. *Also see* specific gaits—CANTER; COUNTER CANTER; DISUNITED CANTER; GALLOP.

ASYMMETRY: In a biological context, any incomplete identity between two sides of the body; or differences in levels of activity between pairs of organs located on either side of the body.

ASYNERGY: Severely reduced or complete lack of coordination among organs which normally act in harmony.

ATAVISM: Reversion to a characteristic(s) typical of very distant ancestors.

ATAXIA: Inability to exhibit taxis due to complete failure or serious irregularity of muscular coordination.

ATOMISM: A theory stating that all conscious states can be rationally analyzed into elementary units. (Early behaviorism and associationism were principally based on atomistic theory.)

ATOMISTIC CATEGORY: Any category that subsumes entities having a narrow range of or identical characteristics.

ATONIA: Lack of tonus in a tissue (e.g., muscular atonia).

ATROPHY: Emaciation of the body or parts of the body of an organism.

ATTACHMENT: *See* specific attachments—PSYCHOLOGICAL; SOCIAL.

ATTACK: A violent, generally sudden, punishing action by one animal toward another.

ATTENTION: Adjustment of sense organs and central nervous system to allow for maximal perception of a particular stimulus. *Also see* specific attention—SELECTIVE.

ATTITUDE: The characteristic manner in which an organism tends to react toward another organism. This term also is used to denote body posture or position, particularly in reference to specific situations, (e.g., defensive attitude, forced attitude, stereotyped attitude).

ATTITUDINAL REFLEXES: A group of reflexes that synchronize the position of the body with the position of the head.

ATTRACTION: The phenomenon of being stimulated to focus attention on, and in some cases to approach, some object, organism, or action because of its inherent characteristics. Antonym: Repulsion.

ATTRACTION SIGNAL: Any vocal, visual, olfactory, or other sign, or combination of such signs, broadcast by an organism to attract other organisms into its spatial proximity.

ATTRACTIVENESS: An attribute of a stimulus that increases the probability of an organism orienting toward and investigating it.

ATTRIBUTE: A basic characteristic of sensation, such as quality, intensity, and duration, that can be analyzed by discrimination tests.

AUDIBILITY RANGE: The range between the lowest and highest frequency of sound perceivable to an individual, group, or species.

AUDIO-: A prefix that indicates a relation to sound.

AUDITION: Perception of sound stimuli (hearing). In mammals and birds, audition is facilitated by the specialized structures and auditory receptors located in the ear.

AUDITORY: Pertaining to the sense of hearing.

AUDITORY NERVE: The part of the eighth cranial nerve which mediates hearing.

AUJESZKY'S DISEASE: A viral disease of the central nervous system affecting cattle, goats, and pigs. Behavioral symptoms, which vary between species, may include pruritus, intense licking, salivation, loss of appetite, paralysis, and in some cases, death. Aujeszky's disease also is called Infectious Bulbar Paralysis or Pseudo-rabies.

AUNT: A female that assists a dam in providing care for young. In natural circumstances, the aunt often is genetically related to the dam (e.g., sister, mature offspring). The term is occasionally also used to refer to females that display epimeletic behavior toward alien offspring or groups of offspring over extended periods of time (e.g., aunt-mare).

AURICLE: The portion of the external ear that surrounds the meatus and extends out from the side of the head. The structure of the auricle is specialized to channel sound waves into the meatus.

AURICLE REFLEX: Movement of the ears in response to auditory stimuli.

AUTO-: Arising from some process within. Not induced by external stimuli.

AUTOCHTHONOUS BEHAVIOR: A term arising out of the concept of action-specific energy. Autochthonous behavior is that which is driven by its own action-specific energy. In this conceptual system vacuum activities could be considered autochthonous behaviors. *Compare*: Allochthonous Behavior.

AUTOMATA: Robot-like mindless devices able to imitate rational actions. Descartes and Cartesian school considered animals to be natural automata.

AUTOMATIC: Acting by itself; spontaneous or involuntary; without conscious control.

AUTONOMIC NERVOUS SYSTEM: A portion of the peripheral nervous system that innervates the viscera, cardiac muscle, blood vessels, smooth muscles, and glands. It was once considered independent of conscious control (autonomous), but now is known not to be completely independent.

AUTOSHAPING: A technique that applies a classical conditioning procedure to an operant response. It involves pairing a neutral stimulus to a reinforcing stimulus, such as food (unconditioned stimulus), that reliably elicits a response. After sufficient pairings, the animal being trained begins to respond (conditioned response) to the formerly neutral stimulus (conditioned stimulus) in the same way as it does to the reinforcing stimulus.

AUTOSOMAL CHROMOSOME: *See* Chromosome.

AUTOSOME: *See* Chromosome.

AVERAGE DAILY GAIN (ADG): The average daily increase in body weight over a given period of time. The average daily gain is used as a measure of growth rate.

AVERSION: A strong or fixed dislike. *Also see* specific aversions—SIGHT; TASTE.

AVERSION THERAPY: Corrective treatment intended to eradicate or inhibit undesirable behavior by associating it with some strongly aversive stimulus.

AVERSIVE STIMULUS: A noxious stimulus that an organism tends to remove or avoid.

AVIARY: A housing system for groups of birds consisting of distinct areas for expression of different activities (e.g., nesting, roosting, feeding, dust-bathing, and water-bathing). Aviaries usually have high ceilings to permit flying by birds.

AVOIDANCE: Prevention or neutralization of aversive stimulation by means of nonapproach, escape, appeasement or subordination, or appropriate operant response.

AWAKE: A state of full consciousness and readiness to perform voluntary activities. Antonym: Sleep.

AWARENESS: The state of being cognizant of some object, entity, or situation.

AXIOLOGY: The study of value theory.

AXIOM: A declaration that does not require proof because it represents self-evident truth or because it is defined implicitly by other inherent axioms.

AXION: The brain and spinal cord.

AXON: A fibrous extension of a nerve cell able to conduct action potentials away from the cell and transmit them to synapsing units.

AZOTURIA: *See* Exertional Myopathy.

B

BABESIOSIS: A parasitic disease of cattle, horses, and sheep caused by various forms of the protozoan, *Babesia*, which destroy red blood cells. Symptoms include high fever, depression, extreme pallor from anemia, and dark-colored urine.

BACKCROSS: To mate a hybrid individual to an individual of one of the genetic stocks from which the parents of the hybrid were derived.

BACKWARD CONDITIONING: A conditioning technique where, contrary to the usual temporal presentation of stimuli, the conditioned stimulus is presented after the unconditioned stimulus. The technique, as such, is not very effective for forming an association between the conditioned and unconditioned stimuli.

BALANCE: Maintenance of an upright posture through appropriate adjustment of muscle tonus. In a more general sense, the term is used to define a state of equilibrium or homeostasis (e.g., nutritional balance, thermal balance).

BALANCED RATION: A food allotment containing all necessary nutrients in proper amounts to adequately nourish an animal at a given age, reproduction status, or level of performance.

BALKING: The refusal of an animal to move from a spot or the disobedience of directive commands from the rider or handler.

BALLOTADE (horse): A dressage exercise of the Spanish High Riding School similar to the Croupade except that the hind legs, instead of being drawn up below the body, are extended backward exposing the hoofs before the horse lands on all four feet simultaneously on the same spot from which it leaped.

BAND (horse): A natural group of horses consisting generally of a stallion, several mares, and several sexually immature individuals of various ages.

BAR BITING: Stereotyped biting, gnawing, or sliding of the mouth on accessible parts (usually metal bars) of an enclosure. This type of behavior occurs relatively often in swine housed in close confinement, particularly farrowing or gestation crates, or on concrete floors without bedding material and with a concentrated diet. Bar biting is considered to be an indicator of boredom, but since its occurrence is increased prior to farrowing, it also may be a component of vacuum nest building or a displacement activity linked to predelivery discomfort. *Also see* Cribbing.

BARED TEETH (horse): Contraction of lips displaying incisor teeth. Bared teeth is a threat signal and is an integral part of aggressive biting.

BARESTHESIA: Sensitivity to pressure (e.g., barometric pressure).

BARK (swine): A short (0.05-0.2 sec), sharp vocalization produced with an open mouth as a single sound or a short series of repeated sounds by a surprised pig. The amplitude of a bark is highest at the beginning of the sound and falls sharply thereafter. The pitch is usually between 2 and 3 kHz. Barks appear to be both threat and alarm calls, having a high arousal effect on other pigs, particularly preweaned young. Other animals react to barking with quick orientation, followed by a brief period of immobility and silence.

BARKER SYNDROME (swine): A rare respiratory disease of pigs. Behavioral symptoms are difficult breathing, repeated barking sounds, and uncoordinated locomotion.

BARN: A building for housing of farm animals.

BAROCEPTOR: A sensory receptor activated by changes in barometric pressure. Synonym: Baroreceptor.

BARORECEPTOR: *See* Baroceptor.

BAROTAXIS: Taxis related to barometric pressure.

BARREN ENVIRONMENT: An environment of very low complexity.

BARRIER, STIMULUS: *See* Stimulus Barrier.

BARROW: A young castrated male pig.

BASAL, BASIC: Defines some fundamental quality or quantity (e.g., basal metabolism, basic activity).

BASAL GANGLIA: The part of the forebrain that is associated with control of body posture.

BASTARD: One whose parentage is not completely known.

BATTERY CAGE: One in a series of identical or similar cages for housing of animals, all equipped with waterers and feeders. Battery cages usually are stacked in several horizontal tiers and most commonly are used for housing of laying hens. The number of hens housed in each cage depends on its size, but generally varies between 3 and 10 birds.

BAWL (cattle): A loud vocalization made when cattle are apparently distressed or disturbed.

BEAK: A hard, external structure of the mouth of a bird enclosing the oral orifice and divided into a fixed upper portion (upper beak) and a mobile lower portion (lower beak). The beak functions as an extension of the jaws, and is used to grasp and manipulate objects and facilitate ingestion of food. It also may be used as a weapon. The beak is innervated through much of its length and is not insensate. *Also see* specific beak—SHOVEL.

BEAK TRIMMING (poultry): Removal of the distal portion of the beak to curtail injuries due to pecking among birds housed in groups. Since the beak is innervated, the ethical validity of beak trimming has been questioned.

BEDDING: Material placed on the floor of a pen to cushion and insulate it, and to absorb wetness. Also, the provision of such material to a pen. *Also see* specific bedding—DEEP.

BEE: *See* specific bees—EXPLORATORY; FORAGER.

BEE DANCE: Rhythmical movement and locomotion performed by bees for the purpose of communication.

BEEFALO: *See* Catalo.

BEE MILK: A secretion of worker bee mandibular glands which is fed to all larvae for a short time, and given to future queens for the full period of their development.

BEHAVIOR: The observable action of a living organism, either instigated by the organism or imposed by external circumstances (e.g., contraction of a muscle, locomotion, vocalization, social interaction, movement as a result of being pushed, etc.). *Also see* specific behaviors—ABERRANT; ABNORMAL; ADAPTIVE; ADJUNCTIVE: AGGRESSIVE; AGONISTIC; ALLELOMIMETIC; ALLOCHTHONOUS; ALTRUISTIC; AMATORY; AMBIVALENT; ANALOGOUS; ANOMALOUS; ANTIPREDATOR; APPETITIVE; APPREHENSIVE; APOPATHETIC; AUTOCHTHONOUS; CARE-GIVING; CARE-SEEKING; CLINICAL; COGNITIVE; COMFORT; COMPETITIVE; COMPROMISE; CONCILIATORY; CONSUMMATORY; CONTACTUAL; CONTAGIOUS; COPULATORY; COOPERATIVE; CRYPTIC: DEFENSIVE; ELIMINATIVE; EMBRYONIC; EPIMELETIC; ESTROUS; ET-EPIMELETIC; EXPLORATORY; EXTRINSIC; FEIGNING; FILIAL; FORAGING; FREEZING; FUNCTIONAL UNIT; HOARDING; HOMOLOGOUS; IMITATIVE; INFANTILE; INGESTIVE; INSTINCTIVE; INTEGUMENTARY; INTENTIONAL; INTRINSIC; INVESTIGATIVE; INVOLUNTARY; MAINTENANCE; MALADAPTIVE; MALADJUSTED; MATERNAL; MORIBUND; NONADAPTIVE; PARENTAL; PATHOGNOMONIC; PERIPARTURIENT; PHORETIC; PLAY; POSTPARTURIENT; PREPARTURIENT; PROCEPTIVE; PUERPERAL; REDIRECTED; REPRODUCTIVE; RITUALIZED; SCHEDULE-INDUCED; SENTINEL; SEXUAL; SHELTER-SEEKING; SOCIAL; SPACING; SPECIES-SPECIFIC; SPONTANEOUS; STEREOTYPED; TERRITORIAL; THIGMOTAXIC; UNCONSCIOUS; VIGILANT; VOLUNTARY.

BEHAVIORAL ACT: *See* Behavioral Action.

BEHAVIORAL ACTION: Any observable behavioral state or event. In a more specific context, the term refers to operationally intrinsic and simple actions which are an integral part of some more complex behavioral pattern.

BEHAVIORAL ANALYSIS: The process of separation of behavioral activities according to their basic features (e.g., cause, function, or history of acquisition).

BEHAVIORAL CATEGORY: A class of behavioral activities, usually based on some essential or fundamental consideration (e.g., learned or instinctive behavior in a classification according to origin; social or territorial behavior in a functional classification).

BEHAVIORAL CYCLE: Any recurrent behavioral sequence consisting of segments which occur in constant order, but which are not necessarily of constant duration. In operant behavior, the most commonly recognized behavioral cycle has a sequence of three phases: appetitive, consummatory and refractory.

BEHAVIORAL DISORDER: Manifestation of behavior that differs from that of a typical healthy animal, thus indicating disease, injury, stress or inability to adjust to the environment. Behavioral disorders may be temporary or chronic.

BEHAVIORAL DISPLAY: Any behavior that has or may have some communicative meaning.

BEHAVIORAL DRIFT: Any unexplained change in the behavior of a given organism or social unit.

BEHAVIORAL EVENT: Any single occurrence of a behavioral state. For purposes of recording, the beginning and end of each such occurrence must be known.

BEHAVIORAL GENETICS: The scientific disciplines concerned with hereditary transmission of behavioral characteristics.

BEHAVIORAL INVENTORY: A list comprising the range of documented behavioral actions performed by an individual, breed, or species. *Compare*: Behavioral Repertoire.

BEHAVIORAL MEASUREMENT: Quantification of a behavior for the purpose of analysis. Also, the numerical expression resulting from such quantification.

BEHAVIORAL MODIFICATION: Any qualitative or quantitative change in an established function or form of behavior.

BEHAVIORAL PARADIGM: A conceptualization that attempts to model the elements of a given behavioral entity.

BEHAVIORAL PATTERN: An organized sequence of behavioral actions having a specific role (courtship displays, nest building, etc.).

BEHAVIORAL REPERTOIRE: The full range of behavioral actions manifested by an individual, breed or species. *Compare*: Behavioral Inventory.

BEHAVIORAL RESILIENCE: The degree to which an animal will reapportion the time budgeted to a particular activity in different circumstances. If the time devoted to this activity varies widely as circumstances change, the activity has high resilience, if the converse is true, then it has low resilience.

BEHAVIORAL RESPONSE: Any behavior performed as a response to a stimulus.

BEHAVIORAL SEGMENT: An arbitrarily defined portion of behavior.

BEHAVIORAL SEQUENCE: Any two or more behavioral segments (actions, patterns, phases, etc.) occurring consecutively in time.

BEHAVIORAL SILENCE: Refers to the understanding that absence of change in an organism's behavior during a potential learning situation allows no conclusion about whether or not anything was learned.

BEHAVIORAL STATE: The behavior of an organism as determined at any given point in time. For purposes of recording behavior, behavioral states are organized into descriptive classes (active-inactive, standing-walking-running, etc.).

BEHAVIORAL SYSTEM: A term used by some scientists to refer to the integrated processes of sensation, cognition, motivation, and motor coordination associated with a given behavioral activity.

BEHAVIORAL TEMPLATE: Used loosely to refer to a genetic predisposition to perform a given behavioral action.

BEHAVIORAL TYPE: A category or class of organisms grouped together on the basis of common behavioral characteristics (frequently designated as personality types in humans).

BEHAVIOR CRITERION: A standard behavior used for comparison with other behaviors, or with different expression intensities of the same behavior.

BEHAVIOR DETERMINANT: A variable that plays a predominant role in elicitation of a given action.

BEHAVIOR DYNAMICS: The development, changeability, and interrelationships of behavioral actions.

BEHAVIORISM: A position, expounded by John B. Watson, that maintained that the subject matter of psychology was behavior, not conscious experience.

BEHAVIOR RATING: Assignment of a rank, score, or mark to a specific observed activity or group of activities.

BEHAVIOR SAMPLING: Observation and recording of the activities of an organism during a prescribed time period for the purpose of obtaining a representative sample of its behavior.

BEHAVIOR SCALE: Any system of categorization whereby a given behavior can be denoted qualitatively or quantitatively. Such systems are appropriate for types of behavior that are scaled over a range of behavior patterns (e.g., threat behavior) or intensities (e.g., speed of locomotion).

BEHAVIOR SHAPING: A training process that uses the conditioning technique of reinforcing any new response that approximates more closely the desired response until the correct behavior is learned.

BEHAVIOR THERAPY: Application of various conditioning techniques to reduce or extinguish undesirable behavior.

BELLOWING (cattle): A general term for loud, repeated vocalizations of cattle. Bellowing may be an indication of excitement or distress (e.g., when expecting feed or water, when separated from group members, when in estrus) or a warning signal. High incidence of bellowing may be associated with nymphomania or disease (e.g., rabies).

BELLY SUCKING: A term applied jointly to navel sucking, prepuce sucking, or juvenile teat sucking.

BENEFIT: An increment in something of value to an animal (e.g., energy). Benefits may accrue from the performance of behavioral actions. In an evolutionary context, benefits are increments in fitness. *Compare*: Cost.

BETA ACTIVITY: *See* Brain Wave Activity.

BIAS: A factor in experimental procedure that introduces systematic error or systematically distorts a set of data.

BIDIRECTIONAL SELECTION: *See* Divergent Selection.

BIGEMINAL: Consisting of two interrelated components, or involving a pair of organisms (e.g., dyad, bigeminal formation).

BIGEMINAL FORMATION: A pair of interacting organisms that may or may not act in concert and be socially bonded.

BIGEMINAL UNIT: *See* Dyad.

BILLING (poultry): Mutual contact of beaks between two birds, often displayed between monogamous partners.

BILL WIPING: Periodic vigorous lateral movements of the head by which a bird rubs its beak on some object. Bill wiping increases in frequency during or after feeding.

BIMATURISM, SEXUAL: *See* Sexual Bimaturism.

BIMODAL: A form of distribution curve which has two maxima.

BINAURAL: Referring to both ears.

BINOCULAR: A device that facilitates vision by two eyes simultaneously; any reference to two eyes.

BINOCULAR VISION: Vision in which images of the same object(s) are projected on the retinas of both eyes simultaneously.

BIOACOUSTICS: The study of sounds produced by animals.

BIOETHICS: A branch of philosophy focusing on ethical issues in biological research and its applications.

BIOLOGICAL CLOCK: A hypothetical mechanism controlling rhythmicity of psychophysiological processes within an organism.

BIOLOGISM: A theory that stresses biological utility as the sole explanatory principle in all aspects of life.

BIOLOGY: The scientific study of living organisms.

BIOMETRY: The study of the statistical analysis of biological data.

BIONOMICS: The study of the relationship between organisms and their environment. Synonym: Ecology.

BIORHYTHM: Any cyclical and generally predictable occurrence of biological phenomena (e.g., estrous cycling, seasonal breeding, circadian hormonal fluctuations, etc.)

BIOSTIMULATION: Stimulation arising from the presence and behavior of peers (e.g., social facilitation, social imitation, Whitten phenomenon, etc.).

BIOTA: The totality of living organisms (both flora and fauna) in a particular geographical area.

BIRTH: Emergence of an organism from the body of its dam. *Also see* specific birth—PREMATURE.

BISEXUALITY: Being sexually attracted to both sexes or possessing characteristics of both sexes.

BITING: Grasping and applying pressure to an object with the teeth or beak to hold, puncture, or shear it. Biting may occur as a component of ingestive behavior, sexual behavior, or aggressive behavior. Habitual aggressive biting is considered a dangerous vice. *Also see* specific biting—BAR; EAR.

BLACKHEAD: A fatal disease occurring mainly in young turkeys (but occasionally also in other species) caused by the protozoan *Histomonas meleagridis*. Behavioral symptoms are drowsiness, gradual inhibition of locomotion, dropped wings and tail, loss of appetite, liquidy greenish excreta, and dark blue or black coloring of featherless parts of the head.

BLANK EXPERIMENT: A preliminary experiment conducted to verify experimental procedure. The results from such an experiment are not incorporated into the data set for final analysis.

BLATT (sheep): A general term for the vocalization of sheep and goats. A medium amplitude sound of variable duration (0.5-2.0 sec) produced with the mouth open. This sound is emitted when an animal becomes separated from its herdmates. In the case of dam-offspring separation, the duration of the dam's blatts is longer than usual. It is thought that short blatts aid herd cohesion and loud, repeated blatts indicate distress.

BLINDNESS: Permanent or temporary inability to see. Blindness is complete (amaurosis) if the organism is unable to perceive any visual stimulus or partial (amblyopia), if only certain visual stimuli are detectable (e.g., color blindness, moon blindness, movement blindness, etc.). Blindness is used occasionally to describe nonvisual perceptual deficiencies (e.g., taste blindness, olfactory blindness). *Also see* specific blindness—PSYCHIC.

BLINDS: Any device that covers an animal's eyes and prevents vision. Blinds are used occasionally for safer handling and casting.

BLIND SPOT: The point where the optic nerve and the central artery of the retina enter the eye chamber. At this spot the retina is unable to respond to visual stimuli.

BLINKER (horse): Flaps attached to a bridle to prevent lateral vision.

BLOAT: A digestive disorder in ruminant animals characterized by excessive accumulation of gases in the rumen. The term also is used to describe generalized distention of the abdomen.

BLOCK, NERVE: *See* Conduction Anesthesia.

BLOCKING: In the context of conditioning, prevention of an association being made between a discriminative cue and a reinforcer by the concurrent presentation of another discriminative cue whose association to the reinforcer already has been established. *Also see* specific blocking—CUE; RAM.

BLOOD TYPING: Serological examination focusing on diagnosis of genetically controlled blood factors. In animals, blood typing is most commonly used for verification of parentage.

BLOOM, IN BLOOM: Refers to cows in early lactation showing promise of good milk yield or, more generally, to animals in good condition.

BLOW (cattle): Forceful expulsion of air through the nasal cavity and nostrils. It is produced during olfactory investigation with the head extended toward the object being investigated. In a louder version, it is apparently a threat signal directed toward a nearby adversary or other aversive stimulus.

BLOW (horse): A sound of approximately 0.15-0.20 sec duration with a broad frequency range. It is emitted often during investigation and is assumed to indicate anxiety.

BOAR: A sexually mature male pig.

BOAR TAINT: Odor in the body tissues of boars caused by muscone.

BODY RUBBER: A simple device, usually a suspended bag, chain, line, or post, against which an animal can rub itself. Body rubbers are occasionally equipped for automatic application of pesticides to control ectoparasites.

BOLTING: Swallowing of food with very little or no chewing action. Bolting can cause digestive problems and typically reduces the efficiency of feed conversion. Habitual bolting is considered a dangerous vice.

BOLTING (horse): Uncontrolled running by the horse which is dangerous to the rider and the horse.

BOND: Any lasting social attachment between individual organisms (e.g., sexual bond between monogamous partners, epimeletic bond between mother and offspring).

BORBORYGMI: Sounds made as a result of peristaltic movement of the intestines.

BOREDOM: A state of weariness and wandering attention, outwardly similar to symptoms of fatigue, caused by lack of environmental complexity. In group-housed animals, boredom may increase the incidence of social vises.

BOSS ANIMAL: "Alpha" animal. (colloquial term)

BOTULISM: A condition arising from eating food contaminated by a strong neurotoxin (botulin) produced by the bacterium *Clostridium botulinum.*

BOUNDARY, DOMINANCE: *See* Dominance Boundary.

BOVINE: A cattle-beast or pertaining to cattle.

BOVINE SPONGIFORM ENCEPHALOPATHY (BSE): A neurological disease of adult cattle, believed to be the bovine equivalent of scrapie disease of sheep, caused by an intake of animal-based proteins containing the scrapie agent. From the onset, the disease slowly progresses over a period of one to six months. The predominant indicators are signs of apprehensive behavior, sensitivity to touch, higher aggressiveness, disorientation, and progressive incoordination. Synonym: Mad Cow Disease.

BOWIE (sheep): A clinical term for leg deformity in lambs. It may lead to severe lameness and inability to walk.

BRACHIAL PLEXUS: A network of nerves in the forelimb, connected to the last three cervical nerves and first one (cattle, sheep, pigs) or two (horses, dogs) thoracic nerves of the spinal cord.

BRADYCARDIA: Abnormally slow heart beat.

BRADYKINESIA: Abnormally slow execution of movements.

BRADYPHAGIA: Abnormally slow eating.

BRADYPNEA: Abnormally slow breathing.

BRADYPRAGIA: Abnormally slow initiation of action.

BRADYTOCIA: Abnormally slow parturition.

BRAIN: The part of the central nervous system located at the anterior end of the spinal cord and situated in the cranium. It is derived from the embryonic prosencephalon (forebrain), mesencephalon (midbrain) and rhombencephalon (hindbrain).

BRAIN LOCALIZATION: The determination of the specific areas of the brain in which particular mental and behavioral processes occur.

BRAIN STEM: The portion of the vertebrate brain comprising the pons, medulla oblongata, and midbrain, and connecting the cerebral hemispheres with the spinal cord.

BRAIN WAVE ACTIVITY: Detectable levels of rhythmical electrical activity of the brain having amplitudes and frequencies corresponding to different levels of arousal. Brain waves are differentiated into alpha waves (8-13Hz), indicative of conscious relaxation; beta waves (14Hz and above), indicative of arousal; delta waves (below 3.5Hz), indicative of SW-sleep; and theta waves (4-7Hz).

BRAMBELL REPORT: The report of the technical committee commissioned to enquire into the welfare of animals kept under intensive livestock husbandry systems, submitted to the British Parliament in 1965. The committee was chaired by F.W.R. Brambell and its formation was largely instigated by the strong public response to the book *Animal Machines,* by Ruth Harrison.

BRANDING: Marking the surface of an organism's body for the purpose of individual identification by causing a permanent and visible scar to form. The most common methods of branding employ a hot iron or a freezing device. Branding destroys melanocytes in the skin and permanently changes the color of hairs growing on the affected areas.

BREAKING-IN (horse): The first stage of training of a young horse.

BREATHING: *See* Respiration.

BREECH PRESENTATION: Fetal presentation in which the posterior enters the birth canal with the hind legs directed toward the head of the fetus.

BREED: Within a species, a genetically closed group of animals having certain heritable characteristics in common. Breeds can be further segregated into strains and lines.

BREEDING: Production of offspring. Also, application of techniques that control or influence the genetic composition of the filial generation. *Also see* specific breeding—ARTIFICIAL; COMBINATION; HAND; NATURAL; NONSEASONAL; RANDOM; SEASONAL.

BREEDING HOBBLES: A restraint device consisting of hobbles fixed around the hocks and firmly tied by a rope to a strong neck collar. Breeding hobbles are used to prevent a mare from kicking a stallion during mating or kicking a person conducting a vaginal or rectal examination.

BREEZE (horse): To run a horse at a moderate speed.

BRIDLING (ducks): A postcopulatory action pattern of male ducks characterized by periodic pulling of the head and neck towards the back interspersed with occasional vocalizations.

BRIGHT BLINDNESS (sheep): A clinical term for a disease affecting sheep that have eaten bracken fern while on pasture. Affected animals have dilated pupils, show poor light reflex, and become blind.

BRIGHTNESS: The amount of light emitted or reflected from the surface of an object.

BRIGHTNESS THRESHOLD: The minimal brightness of an object necessary for an organism to differentiate it from other objects in the visual field.

BROADCAST SIGNAL: A signal that is emitted to convey information to any appropriate individual(s) that may be within range (e.g., crowing by roosters). *Compare*: Directed Signal.

BROILER: A meat type chicken raised for marketing at an early age (approximately 5 to 7 weeks).

BROKEN MOUTH: A condition occurring when a sheep loses its incisor teeth, thus causing difficulties in grazing and feed intake. (colloquial term)

BROKEN PENIS: Morphological abnormality of the penis caused by rupture or tearing of its tissues. The damage may occur as a consequence of a sudden rapid movement (e.g., collapse) of the female during copulation, or misdirection of the penis during strong propulsus.

BRONCHITIS: Inflammation of the bronchial tubes, characterized by increased respiration rate, coughing, nasal discharge, and reduced feed intake. *Also see* specific bronchitis—INFECTIOUS.

BROODINESS: Maternal tendency of birds to hatch eggs in a nest and raise neonates.

BRUCE EFFECT: An increased tendency for abortion in females housed during pregnancy in close confinement with males other than the sire of their fetuses.

BRUCELLOSIS: A bacterial disease of cattle and, to a lesser degree, other farm animals, caused by several forms of *Brucella abortus*. Behavioral symptoms are abortion and occasionally retention of fetal membranes.

BRUSHING (horse): Inappropriate movement of the legs causing the median side of the moving hoof to knock against the median side of the opposite leg.

BRUTE: Any nonhuman animal. The term was used by Descartes who claimed that animals lack language and thus are unable to possess thoughts, and that their behavior can be explained by reference to purely mechanical principles.

BUCK: A sexually mature male goat, rabbit, or deer.

BUCKING (horse): The act of leaping with arched back and lowered head combined with high hindquarters and vigorous kicks. The horse usually lands on stiff forelegs, attempting to dislodge the rider.

BULB, OLFACTORY: *See* Olfactory Bulb.

BULIMIA: Extreme hunger.

BULL: A sexually mature bovine male.

BULLER-RIDER SYNDROME: Frequent mounting of an animal by its peers. This syndrome develops more frequently when the animals are housed in large groups and/or crowded conditions (e.g., in a feedlot) and, if persistent, may cause the death of the mounted animal.

BULL HOOD: *See* Blinds.

BULLING (cattle): Mounting and chin-pressing of peer cows by a cow, usually when the instigator is close to or in estrus. (colloquial term)

BULLY: To harass repeatedly, manifest aggression and despotic behavior toward another individual. Bullying appears more frequently in housing conditions characterized by low environmental complexity, large group sizes, and high population densities, and it is thought that boredom and frustration promote its occurrence.

BUMPING: A traditional technique used to detect pregnancy in cows during the latter stage (last four months) of gestation. It is conducted by placing a hand against the right flank of the abdomen and then applying one or more pushes in an attempt to feel the fetus. The accuracy of this technique increases with advancing stage of pregnancy and experience of the tester.

BUNTING: *See* Butting.

BUNT ORDER: Social rank order in pigs, sheep, goats, and cattle. (colloquial term)

BURROWING: To dig into and make passageways through some material (e.g., a rabbit burrowing through soil or straw).

BUTTING: Hitting an object or another organism with the forehead or horns. Butting is commonly displayed during aggression or as displacement behavior when directed toward some object other than an adversary. A mild form of butting may also be observed as a part of courtship activities. Butting during nursing (nursing butts) conducted by a suckling offspring is most common in bovines.

CACKLING (chicken): A low frequency vocalization consisting of a relatively long (up to 4 sec) sequence of uneven duration sounds, often produced after egg laying.

CADAVER: A dead body.

CAESAREAN DELIVERY: Surgical removal of a fetus from the uterus usually at the approximate time when parturition would occur normally.

CAGE: An enclosure made of metal, plastic, or wooden bars or wire for housing or transport of animals. *Also see* specific cages—BATTERY; COLONY; GETAWAY.

CAGE LAYER FATIGUE: Leg paralysis and bone fragility occurring in laying hens housed in wire floor cages, assumed to be caused by lack of exercise and calcium depletion. (colloquial term)

CALF: In some mammalian species (such as cattle, elephants, whales, etc.), an animal before it reaches puberty.

CALL: *See* specific calls—ABANDONMENT; ALARM; DISTRESS; SEPARATION; WARNING.

CALLS, SEQUESTRATION: *See* Sequestration Calls.

CALVING (cattle): Natural parturition in cattle.

CALVING EASE: Delivery ease in cattle (e.g., no-pull, easy-pull, hard-pull, surgical).

CALVING INTERVAL: Time interval between two consecutive calvings.

CAMOUFLAGE: *See* Cryptic Coloration.

CAMPAGNE (horse): Elementary dressage training, consisting of collection, neck arching, weight shifting on hindlegs, riding in a straight line, turning, and lateral movements.

CANINE: A member of the family Canidae, or pertaining to dogs, coyotes, wolves, etc.

CANNIBALISM: The practice of consuming the tissue of conspecifics. This term also is used colloquially for the killing or serious injury of conspecifics by biting or pecking. *Compare*: Kronism.

CANTER: A naturally developed three-beat gait. The sequence of hoof beats, with the right foreleg as the leading leg, is left hind, right hind together with left fore, then right fore followed by complete suspension in the air. The diagonal pair changes with change of the leading leg. When taking a curve, the leading leg is the one on the inside of the curve. *Compare*: Disunited Canter; Counter Canter.

CAP: A visible swelling on the joints of animal limbs, e.g., capped elbow, capped hock, capped knee.

CAPON: A castrated male chicken.

CAPRILLIC: Referring to goats, particularly goat odor.

CAPRINE: A goat or pertaining to goats.

CAPRIOLE (horse): A dressage exercise of the Spanish High Riding School. The horse leaps into a ballotade, kicks powerfully with both hind legs to reach a horizontal level, then lands on all four hoofs simultaneously near the place the leap was initiated.

CARDIAC: Pertaining to the heart (e.g., cardiac muscle).

CARDIAC ARREST: Cessation of rhythmic contraction of the heart muscle. One of the most common causes of cardiac arrest is severe hypoxia of the heart, which reduces excitability of its muscle fibers.

CARDIAC REFLEXES: A complex set of reflexes that, together, comprise the action of the cardiac muscle.

CARE: *See* specific care—MATERNAL; PATERNAL.

CARE-GIVING BEHAVIOR: *See* Epimeletic Behavior.

CARE-SEEKING BEHAVIOR: *See* Et-epimeletic Behavior.

CARNIVOROUS: Pertaining to meat-eating species.

CARTESIAN DOCTRINE: A doctrine expounded by the philosopher Descartes, declaring animals to be natural automata whose behavioral repertoire can be explained fully by mechanical principles. This doctrine claims that animals do not have the ability to develop thoughts on the grounds that they do not possess language and therefore are incapable of abstraction. Cartesian doctrine aroused serious objections from other philosophers (e.g., Voltaire, Hume) who argued that animals are capable of inferential reasoning to some degree.

CAST: To cause an animal to fall to the ground. Casting can be done physically or with drugs and can be used to restrain animals for certain types of examination or treatment.

CASTE: An exclusive social class within some larger social structure.

CASTING, MEDICAMENTOUS: *See* Medicamentous Casting.

CASTING HARNESS: A harness consisting of a strong leather girth tied around an animal's chest, and a rope that passes through both a metal ring on the girth and hobbles on the pasterns of all four legs, being tied to the last hobble. Pulling the rope forces the animal into recumbency, and the legs can be tied together.

CASTRATION: Surgical removal of or rendering nonfunctional an animal's gonads. *Also see* specific castration—PSYCHOLOGICAL.

CATALO: Offspring of a cross between domestic cattle and bison.

CATARRH, NASAL: *See* Rhinitis.

CATATONIA: Muscular rigidity that temporarily suppresses locomotion. Organisms in a catatonic state seem to be fully aware of their surroundings and able to visually monitor moving objects. Catatonia is considered the consequence of psychological disorder.

CATEGORY, ATOMISTIC: *See* Atomistic Category.

CATEGORY, MOLAR: *See* Molar Category.

CATTLE: Domesticated bovine animals.

CATTLE LEADER: *See* Nose Lead.

CAUDAL: Pertaining to or located in the direction of the posterior end of an animal's body.

CAUSAL ANALYSIS: An analysis focusing on the factors that determine or influence a given behavioral action or behavioral pattern.

CAUSAL FACTOR: Any identifiable entity or circumstance in the internal or external environment of an organism presumed to be primarily responsible for its individual psychological states and behavioral activities.

CAVALCADE (horse): A procession of ridden horses or horsedrawn carriages.

CAVITY, TYMPANIC: *See* Tympanic Cavity.

CENTER: The middle point of a body; a conglomerate of integrated neurons that fulfil specific perceptual functions (e.g., auditory, gustatory, olfactory, visual center) or control and coordinate specific activities (e.g., respiratory, thermoregulatory, vocalization, micturition, defecation, vomiting center).

C

CENTRAL NERVOUS SYSTEM (CNS): The brain and spinal cord.

CENTRAL TENDENCY: A disposition to gravitate toward the center of a ranking scale.

CEPHALO-, CEPHAL-: Relating to the head.

CEPTOR: *See* Receptor.

CER: *See* Conditioned Emotional Response.

CEREBELLAR SYNDROME: A clinical term for a condition caused by disease of or damage to the cerebellum. Behavioral symptoms consist of a broad variety of actions indicative of impaired motor coordination. Cerebellar syndrome is characterized by compulsive movements, abnormal postures, erratic gait, inability to maintain posture or balance, nystagmus, and dysmetria.

CEREBELLUM: The part of the hindbrain consisting of the median ridge and two lateral lobes. It is associated with co-ordination of muscular action to maintain balance and execute movements.

CEREBRAL CORTEX: The outer layer of the cerebral hemispheres consisting of gray matter. The cerebral cortex is the center of cerebration.

CEREBRAL HEMISPHERES: A pair of spherical structures of the brain separated by a deep longitudinal cleft and interconnected at their bases by the corpus callosum. The cerebral hemispheres, which develop from the embryonic telencephalon, are subdivided into sections and lobes, and are covered on the surface by a layer of cortex. They function as centers for integration of diverse afferent sensory input and control of voluntary action.

CEREBRAL SYNDROME: A clinical term referring to a condition caused by disease of or damage to the cerebrum. Behavioral symptoms indicate impaired perceptual ability of the affected animal and low level of awareness of its surroundings. Cerebral syndrome often is accompanied by staggering gait and stereotypic head movements.

CEREBRATION: A process of cerebral analysis that integrates sensory input with information stored in memory to select appropriate behavior.

CEREBROCORTICAL NECROSIS: A disease, also called polioencephalomalacia, characterized by necrosis of the cerebral cortex and cerebral edema. Behavioral indicators are head pressing and blindness. Cerebrocortical necrosis occurs occasionally in cattle and sheep.

CEREBROSPINAL SYSTEM: *See* Central Nervous System.

CEREBRUM: The two cerebral hemispheres connected by the corpus callosum. *Compare*: Cerebral Hemispheres.

CEREMONY: A synchronized display involving two or more individuals characterized by a standardized sequence of behavioral actions. Visual, auditory, and tactile modes of communication usually are part of ceremonies. *Compare*: Ritualized Behavior.

CEREMONY, TRIUMPH: *See* Triumph Ceremony.

CHAFING: Any integumentary injury caused by improper harness fitting (e.g., curb-gall, girth-gall, etc.).

CHAIN: *See* specific chains—ACTION; COURTSHIP.

CHAIN REFLEX: An uninterrupted sequence of reflexes activated by a specific stimulus. Each subsequent reflex is induced by its predecessor.

CHAMPING: Repetitious, strong opening and closing action of the mouth which produces sounds when the teeth hit together. Champing in swine may be a threat signal, but also is performed by boars during courtship and mating.

CHANCE: In a biological sense, an event that appears to be random, accidental or unpredictable.

CHANT-DE-COEUR (swine): *See* Courtship Grunts.

CHARACTER: The combination of persistent habits, reactions, and other individual traits of an organism.

CHARGING: A threatening lunge toward or attack on a human by an animal. Charging most commonly refers to an aggressive action of a stallion, bull, boar, or ram towards an attendant.

CHEMORECEPTOR: A sensory receptor activated by chemical stimulation (e.g., gustatory, olfactory receptors).

CHEMOTAXIS: Taxis in response to chemical substances.

CHESTNUT (horse): A hornlike growth on the medial side of the leg. On the forelegs, chestnuts are located above the knee, on the hind legs, below the hocks. The term chestnut also may refer to horse coat color.

CHEW: To masticate food or other substances.

CHEWING: *See* specific chewing—HAIR; SHAM.

CHIASMA, OPTIC: *See* Optic Chiasma.

CHICK: A young chicken.

CHIMERA: An organism whose body combines portions of different zygotes or embryos, or contains tissue cells of another organism.

CHIN MARKER: A device attached to the chin of a marker animal that leaves color marking(s) on the body of an animal subjected to mounting or chin-pressing. Chin markers are used for estrus detection.

CHIN PRESSING: Positioning of the chin on the dorsal area of a peer. Chin pressing is performed most frequently by bovine males during precopulatory behavior but also by estrus and periestrous cows. Chin pressing is usually oriented toward the withers and the rump. Chin pressing also may be performed as chin rubbing.

CHIN RUBBING: *See* Chin Pressing.

CHIRRUP (swine): A short vocalization sound (0.1-0.2 sec) of young piglets characterized by rapidly varying pitch. It occurs during the first two months of life.

C

CHOICE: Selection from two or more alternatives.

CHOICE POINT: A place in a discrimination apparatus where the organism is expected to choose the direction in which to proceed.

CHOKING: Interruption of respiration accompanied by coughs and attempts to vomit caused by obstruction of the pharynx by foreign objects or food.

CHOLERA, FOWL: *See* Fowl Cholera.

CHOLERA, HOG: *See* Hog Cholera.

CHOLERIC: Having an irritable temperament. The term derives from classical categorization of human character types. Antonym: Phlegmatic.

CHOREA: Irregular and strong spasmodic involuntary movement.

CHROMATIC: Pertaining to hue, or colors in general.

CHROMATOTAXIS: Taxis in response to color.

CHROMOSOME: A structure, composed of protein and DNA, occurring within cellular nuclei and containing a specific set of genes. An animal carries a standardized group of chromosomes in each cell containing a full complement of genes characteristic of the species to which the animal belongs. In mammals and birds, either of the two types of chromosome that are involved in sex determination are called sex chromosomes. Autosomal chromosomes (or autosomes) are those that are not involved in sex determination.

CHRONIC: Persisting over a long time.

CHRONIC DISEASE: Any disease that progresses very slowly. *Compare*: Acute Disease.

CHRONIC LAMENESS: Lameness that persists for an extended period of time. Such lameness may be caused by long-lasting injury or permanent deformity.

CHRONOLOGICAL AGE: The elapsed period since the birth of a given individual. *Compare*: Physiological Age.

CHUTE: A narrow passage designed to direct and regulate the movement of animals. Chutes are used to facilitate loading, unloading, medical treatment, weighing, inspection of animals, etc.

CHYLE: Partly digested food passing through the small intestine.

CHYME: Partly digested food passing from the stomach into the small intestine.

CIRCADIAN: Referring to cyclic rhythmicity corresponding closely to a 24-hour interval.

CIRCADIAN TIMER: A hypothetical psychophysiological mechanism that controls the timing of biological phenomena (e.g., behavioral actions) according to a cyclic rhythm of about 24 hours.

CIRCALUNAR: Referring to cyclic rhythmicity corresponding closely to the lunar interval, i.e., 29.5 days.

CIRCANNUAL: Referring to cyclic rhythmicity corresponding closely to a yearly interval.

CIRCLE (horse): A dressage maneuver in which the horse follows a circle larger than 6 m in diameter.

CIRCLING: Any repeated circular locomotion. Circling may be stereotypic behavior or a symptom of a neural disease (e.g., meningoencephalitis, listeriosis, etc.). Circling can also be induced by conditioning.

CIRCLING DISPLAY: Circular movement of sexual partners in a synchronized head-to-back position; considered to be a courtship activity.

CIRCUIT, NEURAL: *See* Neural Circuit.

CIRCULAR REACTION: Any response that stimulates its repetition.

CIRCUMDUCTION: Movement of a body (or portion of a body) such that it describes a cone, i.e., one end remains fixed in space and the other end travels through a circle.

CIRCUS MOVEMENT: *See* Circling.

CLAN: A social unit consisting of number of families which can trace their ancestry to a common progenitor.

CLASSICAL CONDITIONING: A type of conditioning in which an originally neutral stimulus (the conditioned stimulus) is repeatedly paired in temporal sequence with another stimulus (the unconditioned stimulus) which reliably elicits a response (the unconditioned response). After a sufficient number of such pairing trials, the conditioned stimulus starts to elicit a response (the conditioned response) similar to the original unconditioned response. Synonym: Pavlovian Conditioning, Respondent Conditioning.

CLASSIFICATORY SCALE: *See* Nominal Scale.

CLAW, SLIPPER: *See* Slipper Claw.

CLEANSING: Afterbirth. (colloquial term)

CLIFF, VISUAL: *See* Visual Cliff.

CLIMATE: The spectrum of weather conditions pertaining to a given area or locality.

CLINCH POSITION: Body position of two fighting adversaries, characterized by mutual insertion of the head between the hind legs of the opponent. The clinch position occurs more frequently when the duration of the fight is long and is assumed to provide each animal with an opportunity to rest while in partial control of its opponent.

CLINIC: An establishment for diagnosis and treatment of various disorders, including behavioral disorders.

CLINICAL BEHAVIOR: Behavior indicative of disease or injury.

CLIQUE: A group within a group that operates as a cooperative unit to gain specific advantages for its members. The formation and exploitation of clique affiliation is occasionally called "pactive behavior."

CLOCK, BIOLOGICAL: *See* Biological Clock.

CLONUS: Rapid, successive contraction and relaxation of muscles.

CLOSEBREEDING: Sexual reproduction in which gametes originate from an-

C

imals that have a close genetic relationship (e.g., brother-sister or parent-offspring). Closebreeding is considered an intense form of inbreeding.

CLOSED POPULATION: A social unit into which introduction of individuals from other social units for breeding purposes is not permitted. Smaller social units with the same restriction are called closed herd or closed flock.

CLUTCH: A set of eggs laid on consecutive days.

CO-ADAPTATION: Mutual adaptation between two or more species.

COAT: Pelage of mammals.

COAT MOUTHING: *See* Hair Chewing.

COCCIDIOSIS: A disease caused by *Coccidium zurnii* or various forms of *Eimeria*. Behavioral symptoms are severe diarrhea and rapidly progressive weakening, often leading to paralysis and death.

COCHLEA: The snail shaped part of the inner ear; a spiral form.

COCK: A male chicken more than one year old. Synonym: Rooster.

COCKEREL: A male chicken less than one year old.

COGNITION: A process of perception, reasoning, and development of expectations.

COGNITIVE BEHAVIOR: Behavior that results from a process of reasoning.

COGNITIVE MAP: A hypothetically defined conceptualization of the spatial arrangement of the environment in the memory of an organism.

COHESION: Intentional and mutual maintenance of close proximity among two or more conspecifics for some sustained period of time (e.g., social cohesion among individuals in a herd).

COINCIDENCE: Joint occurrence of two or more independent events.

COITAL LOCK: Powerful contraction of the vulva around the bulb of an inserted penis which temporarily prevents separation of copulating partners. Coital lock is common in copulating dogs.

COITUS: Intromission of the penis into the vagina and per vaginal connection between male and female. Coitus facilitates natural transfer of ejaculated semen from the male reproductive tract to the female reproductive tract.

COITUS INCOMPLETE, COITUS INTERRUPTUS: Coitus which is terminated before semen ejaculation.

COLIBACILLOSIS: A disease caused by pathogenic strains of *Escherichia coli*. Behavioral symptoms are defecation of white, pasty, or watery excreta; shivering; raised hair coat or plumage; and rapid deterioration of body constitution.

COLIC SYMPTOMS: Behavioral symptoms of colic (abdominal pain) are repeated lying down, rolling on the ground, lying on the back, and sitting in a dog-like position.

COLONY: A long-lasting, naturally forming, socially coordinated group of animals in which the individuality of group members is subordinated, to a greater or lesser extent, to the continued existence and activity of the colony.

COLONY CAGE: A cage equipped with feeder and waterer(s) for housing groups of animals. The number of animals housed in each cage depends on its size, but usually varies between 10 and 40. Colony cages are used primarily for growing poultry.

COLONY FISSION: Multiplication of colonies through separation and departure of one or more reproductive subcolonies.

COLOR: A quality of a substance as determined by the wavelength of light (hue) reflected from its surface. *Also see* specific color—PRIMARY; SEMANTIC.

COLOR ADAPTATION: Decrease in an organism's threshold of sensitivity to hue.

COLORATION: Coloring and arrangement of color patterns on an organism's surface.

COLORATION, CRYPTIC: *See* Cryptic Coloration.

COLOR BLINDNESS: Inability to perceive chromatic differences. The degree varies from inability to discriminate between some hues (e.g., red and green) of the same brightness to complete insensitivity to any hue.

COLORIMETER: An instrument that measures colors by comparing them to a standard color mixture.

COLOSTRUM: A fluid, rich in minerals and immunoglobulins, produced in the mammary gland during the periparturient period. In many species of mammal (cows, pigs, horses, etc.), it is essential for neonates to ingest colostrum very soon after parturition to develop resistance to disease.

COLT: A young male horse.

COMA: A state of unconsciousness during which most behavior and reflexes are inhibited.

COMBAT: *See* Fight.

COMBINATION BREEDING: A system of reproduction designed to utilize specific combining ability between parental genotypes to produce offspring conforming to given production objectives.

COMFORT: A state of ease and freedom from noxious stimulation. Subjective interpretation of comfort by an individual depends on many factors, such as level of sentience, age, previous experience, health, etc.

COMFORT BEHAVIOR: *See* Comfort Movement.

COMFORT MOVEMENT: Any movement performed to temporarily relieve muscular tension or integumentary irritation (e.g., grooming, preening, scratching, wing flapping, shaking, stretching, etc.).

COMFORT ZONE: The range of environmental conditions in which an organism experiences comfort. Comfort zones usually are defined in terms of specific attributes of the environment, such as thermal comfort zone, spatial comfort zone, respiratory comfort zone, etc.

COMMENSALISM: A form of interspecies relationship in which one species benefits while the other remains unaffected.

C

COMMUNICATION: A process whereby one system or unit, e.g., an organism, neurophysiological structure, gene, etc., influences another through transmission of a signal(s).

COMPARATIVE ETHOLOGY: A branch of ethology that focuses on the detection of behavioral similarities or differences among species.

COMPARATIVE PSYCHOLOGY: A branch of psychology that focuses on the detection of psychological similarities or differences among species.

COMPENSATION: Counterbalancing of a defect in body structure or in surroundings by behavioral adjustment or alternate behavior.

COMPETITION: The attempt to gain adequate access to and use of a resource(s) (e.g., food, water, shelter, space, sexual partner, high social rank) to satisfy animal needs or desires when the resource availability is not sufficient to allow ad libitum consumption by all individuals. *Also see* specific competition—MOTIVATIONAL.

COMPETITIVE BEHAVIOR: Behavior manifested to render an animal effective, relative to other animals, in the acquisition of some resource (e.g., aggression, mating calls and displays, foraging strategy, territoriality).

COMPLETE FEED: A feed that has been processed to incorporate all required nutrients into one mixture.

COMPLETE LEARNING METHOD: A method of studying rate of learning by determining the number of trials necessary to fully master a given learned response. Synonym: Complete Mastering Method.

COMPLEX HIERARCHY: A type of social hierarchy in which dominance-subordinance ranking includes two or more hierarchy loops. *Compare*: Linear-Tending Hierarchy.

COMPLEXITY, ENVIRONMENTAL: *See* Environmental Complexity.

COMPOSITE SIGNAL: A signal whose meaning is determined by the combination of the simpler signals comprising it. For example, the signal by which a sow indicates preparedness to nurse offspring consists of nursing grunts coupled with positional availability of teats.

COMPOSITION, GROUP: *See* Group Composition.

COMPOUND STIMULUS: A stimulus in which experimentally separable components occur simultaneously.

COMPROMISE BEHAVIOR: Behavior that is expressive of motivational conflict. The animal may manifest actions simultaneously that are indicative of different motivations, or may alternate quickly between actions consistent with different motivations.

COMPULSION: An irresistible impulse or tendency to perform an act.

COMPULSIVE MOVEMENT: A clinical term for stereotypic behavior, e.g., circling, body twisting, pressing against a wall or pulling on a halter.

COMPULSIVE ROLLING: Serious disturbance of postural reflexes characterized by excessive wallowing on the ground and difficulty or inability to stand up. Compulsive rolling is an indicator of brain damage.

CONCENTRATE: Any food for livestock that is high in energy or protein and low in crude fiber.

CONCENTRATION: In a behavioral sense, a high intensity of attention.

CONCEPTION: Fertilization of an ovum.

CONCEPTUALIZATION: Formulation of object hypotheses about events or associations between events based on abstractions and generalizations requiring a complex cerebration process.

CONCILIATORY BEHAVIOR: Behavior directed toward other organisms that tends to promote favorable social relationships. Subordination signals, greeting signals, and appeasement signals are the most common displays of conciliatory behavior.

CONCRETE: Existing or occurring in itself, not merely as an idea or hypothetical construct.

CONDITION: The general health and soundness of an animal, or the amount of flesh or finish on it.

CONDITIONAL STIMULUS: *See* Conditioned Stimulus

CONDITIONED EMOTIONAL RESPONSE (CER): A response that develops when a subject associates a certain situation with some strong experience. CER may develop when an experimental animal associates a testing room or apparatus with either a traumatic or a pleasant experience. Later, the animal will manifest signs of emotionality before entering the room or its responsiveness in the room will be affected.

CONDITIONED FEAR: Any fear attained through a conditioning process. Conditioned fear is an example of a conditioned emotional response.

CONDITIONED INHIBITION: Partial or complete suppression of a conditioned response resulting from pairing of a conditioned stimulus with some other stimulus in the absence of the original unconditioned stimulus.

CONDITIONED REINFORCER: A neutral stimulus that has acquired the functional property of a reinforcer by being paired in time with a primary reinforcer.

CONDITIONED RESPONSE: The response evoked by the conditioned stimulus after conditioning has taken place.

CONDITIONED STIMULUS: A stimulus that is paired in temporal sequence with an unconditioned stimulus and, thus, subsequently acquires the capacity to evoke the identical response as that elicited by the unconditioned stimulus. For example, a sound after being paired with the presentation of food for a period of time, can become a conditioned stimulus capable of eliciting salivation.

CONDITIONING: Formation of an association to an originally neutral stimulus through reinforcement. Development of such an association is influenced by motivational state and previously established associations. *Also* *see* specific conditioning—BACKWARD; CLASSICAL; DELAYED; EXPERIMENTAL; INSTRUMENTAL; INTEROCEPTIVE; OPERANT; PAVLOVIAN; PSEUDO-; RESPONDENT; SECOND-ORDER; TEMPORAL; TRACE.

CONDUCTION: Transmission of energy (e.g., heat, action potentials, etc.) through some medium.

CONDUCTION ANESTHESIA: Diminishment or full inhibition of conductivity of action potentials along a neuron by injection of anesthetic into perineural tissue.

CONE: A photo-sensitive cell in the retina that functions as a receptor for color vision. *Compare:* Rod.

CONFIGURATIONISM: Interpretation theory which accentuates interdependence of behavioral measurements and rejects conclusions based on a single parameter or a narrow set of parameters.

CONFINEMENT HOUSING: A housing system where animals are tethered or otherwise restricted in ambulation.

CONFLICT: A psychological state which occurs when an animal is simultaneously motivated to manifest two or more mutually incompatible responses (e.g., approach-approach conflict, when the organism has to make a choice between two or more attractive stimuli; approach-avoidance conflict when the organism is in a situation where attraction and aversion are inseparably linked with the stimulus; avoidance-avoidance conflict, when the organism has to make a choice between two or more aversive stimuli). Compromise or displacement behavior may occur when the conflicting motivations are equal in strength.

CONFLICT THEORY OF DISPLAY: A theory suggesting that many of the action patterns in behavioral display sequences have become ritualized from ambivalent, compromise, or displacement actions manifested when an animal experiences approach-avoidance conflict during social encounters.

CONGENITAL: Acquired during development prior to birth or hatching.

CONGRESS: Voluntary assembly of two or more conspecifics for a specific purpose, e.g., reproduction (sexual congress), preparation for migration (migratory congress), etc.

CONNATE: Appearing at birth or shortly thereafter.

CONNECTION: In a neuropsychological sense, a linkage between receptors and effectors.

CONSCIOUSNESS: A state of awareness of and responsiveness to information derived from the senses. Consciousness also may refer to a state of self-awareness.

CONSCIOUS PROCESS: Refers to a process of which the organism is fully aware.

CONSEQUENTIALISM: *See* Consequentialist Moral Theories.

CONSEQUENTIALIST MORAL THEORIES: Theories that claim that moral rightness or wrongness of human actions can only be assessed from the consequences of such actions. Utilitarianism and Rational Egoism are examples of consequentionalist moral theories. *Compare*: Deontological Moral Theories.

CONSERVATION: Any activity having the objective of saving, protecting, and/or limiting use of some resource or entity (e.g., body energy, food reserves, or a species of wildlife).

CONSOLIDATION: The process by which information becomes established in long-term memory. Consolidation is thought to involve changes in structure and function of synaptic connections.

CONSPECIFIC: Pertaining to individuals belonging to the same species.

CONSTIPATION: Infrequent or difficult elimination of feces.

CONSTRUCT, HYPOTHETICAL: *See* Hypothetical Construct.

CONSUMMATORY ACT: An act by which an organism satisfies its interests.

CONSUMMATORY BEHAVIOR: Behavior consisting of consummatory responses manifested during the consummatory phase of an operant behavioral cycle.

CONSUMMATORY PHASE: The middle phase of an operant behavioral cycle, preceded by an appetitive phase and followed by a refractory phase. *Compare*: Behavioral Cycle.

CONSUMMATORY RESPONSE: A behavioral response by which an organism attains a goal.

CONTACT RECEPTOR: A sensory receptor which requires mechanical contact with the stimulus to be activated. *Compare*: Teloreceptor.

CONTACT SENSATION: Sensation of touch.

CONTACT SPECIES: Species whose members seek a lifestyle based on close social grouping with peers.

CONTACTUAL BEHAVIOR: Behavior characterized by bodily contact.

CONTAGIOUS BEHAVIOR: *See* Allelomimetic Behavior.

CONTENTMENT: An emotional state resulting from pleasing satiety.

CONTIGUITY: Nonrandom occurrence of events or objects close in time and space. When repeated, such occurrences lead to associative interpretation of these events or objects.

CONTINUOUS DISTRIBUTION: A frequency distribution consisting of divisible units (e.g., time, speed, weight, volume, etc.).

CONTINUOUS REINFORCEMENT SCHEDULE: A reinforcement schedule in which the reinforcing stimulus is presented after every correct response. The Continuous Reinforcement Schedule can be considered a special case of the Fixed Ratio Reinforcement Schedule when the ratio of response to reinforcement is 1:1. *Compare*: Intermittent Reinforcement Schedule.

CONTRACTARIANISM: An ethical theory which implies that moral obligations derive ultimately from agreements, contracts, or consented codes.

CONTROL, YOKED: *See* Yoked Control.

CONTROLLED FEEDING: Any feeding program where the delivery of feed and/or the amount of feed is controlled by humans.

CONTROLLED MOLTING: Molting induced by humans.

CONTROL SYSTEMS THEORY: An approach to the analysis of the control of behavior in which theoretical concepts derived from engineering are used to develop predictive models. Such models are constructed according to quantitative assumptions and by means of computer simulation. The reliability of the construct is compared to experimental data obtained from animals in real situations.

CONVERGENCE: An evolutionary process in which two species independently develop some similar characteristics.

CONVIVIUM: Any social system existing in a given time. Social environment.

CONVULSION: Violent muscular contractions affecting all or part of the body. Convulsions may be a symptom of encephalomyelitis, brain edema, tumor, epilepsy, lesion, or parasitic injury to the nervous system, or may be associated with hysteria, high fever, or parturition.

COOLIDGE EFFECT: A phenomenon which occurs when the copulatory activity of a male increases upon his exposure to a different female.

COOLING RESPONSE: Any response, such as panting, feather sleeking, wing raising, wallowing in a wet area, etc., performed to dissipate body heat.

COOPERATIVE BEHAVIOR: Behavior by which two or more organisms work together to achieve a goal (e.g., coordination of behavior during mating, territorial defense, or feeding of offspring).

COORDINATION: Refers to the ability of motor centers to control contraction and relaxation of muscles necessary for efficient, harmonious execution of required movements. The term also is used to indicate synchronization of behavior among two or more individual organisms (e.g., pulling in horses, group defense behavior).

COPE: An animal's ability to deal with the contingencies of its surroundings without suffering harm.

COPIOPIA: Eye fatigue resulting from overwork of the eyes or from improper illumination.

COPROPHAGIA: Eating of own excrements or excrements of conspecifics.

COPULATION: *See* Coitus.

COPULATION, FORCED: *See* Forced Copulation.

COPULATORY BEHAVIOR: Behavioral actions directly involved in copulation. These include cooperative posture of a female and mounting, intromission, thrusting, ejaculation, and dismounting of a male. Copulatory behavior is preceded by precopulatory behavior (e.g., search for sexual partner, courtship) and followed by postcopulatory behavior (e.g., grooming, resting, etc.).

CORNEAL REFLEX: Rapid closure of the eyelids when the cornea is irritated. Also called blink reflex or lid reflex.

CORNERING: A behavioral phenomenon in which an organism cannot escape a threatening adversary without instigating an attack on the adversary.

CORNERING (chicken): A postural courtship display of males characterized by running (preferably into a corner), stamping feet, and lowering of the body, often in conjunction with tidbitting calls.

CORNERING, MOTIVATIONAL: *See* Motivational Cornering.

CORPUS LUTEUM: A temporary structure formed on the ovary after ovulation by proliferation of cells in the follicular cavity. The corpus luteum secretes progesterone, which prevents estrus cycling and maintains pregnancy. If conception does not occur or if pregnancy is not maintained, the corpus luteum gradually disappears.

CORRELATION: The degree to which variables change together. The correlation coefficient ranges from -1 to +1 and is used to characterize genetic, environmental, and phenotypic associations.

CORTEX, CEREBRAL: *See* Cerebral Cortex.

CORTICOSPINAL TRACT: A neural tract that begins in the motor cortex and extends through the midbrain and brain stem and into the spinal cord. In mammals, the voluntary control of movement is mediated predominantly through this tract.

CORTICOSTEROID: One of several hormones secreted by the adrenal cortex that control carbohydrate metabolism and electrolyte balance. Corticosteroids are used in treatment of severely stressed or shocked animals. Intravenous application of corticosteroids may induce abortion.

CORTICOTROPIC RELEASING FACTOR (CRF): A polypeptide synthesized by neurosecretory cells of the hypothalamus. Its release increases in response to stress. CRF appears to control ACTH release.

CORYZA: *See* Rhinitis.

COSSET: A lamb raised without its dam.

COST: A decrement in something of value to an animal (e.g., energy). Costs are incurred in the performance of behavioral actions and must be weighed against the benefits accruing from such actions. In an evolutionary context, costs are decrements in fitness. *Compare*: Benefit.

COST-BENEFIT ANALYSIS: A hypothetical component of the cognitive process whereby an animal decides among alternative courses of action. It is proposed that an assessment of the costs potentially incurred by given behavioral actions and the benefits potentially resulting from the same actions is made by the animal making the decision. A well-adapted animal would be expected to make a choice that would maximize the benefit relative to the cost.

C

COTERIE: A social unit in which members cooperate to achieve a common goal, such as the defense of their territory against intruders.

COUGH: *See* Cough Reflex.

COUGH REFLEX: A reflex mediated by the medulla oblongata through the vagus nerve, instigated by the presence of obstructing matter in, or irritation of, the respiratory pathways. Coughing is performed as one or several relatively powerful bursts of air expelled from the respiratory tract to remove the obstruction or alleviate the irritation. Repeated coughing when not engaged in eating may be an indication of respiratory disease.

COUNTER CANTER (horse): The counter canter occurs when the horse takes a right curve with the left foreleg being the leading leg, or vice versa. *Compare*: Canter.

COUNTERCONDITIONING: Conditioning to change the valence of a conditioned stimulus (CS) or unconditioned stimulus (UCS) so that an animal develops a response to the stimulus that is mutually exclusive to the response it originally manifested. The original CS or UCS in effect becomes perceived as a qualitatively different CS that elicits a different conditioned response. Counterconditioning generally is carried out in training situations to change an animal's response to a stimulus from an undesirable action to one that is more desirable. The phenomenon also may occur without human mediation, e.g., psychological castration, taste aversion.

COUPLING (swine): A term referring to mating in swine.

COURBETTE (horse): A dressage exercise of the Spanish High Riding School in which the horse makes several forward leaps in the Pessage posture.

COURTSHIP: Specialized behavior patterns that occur as preliminaries to mating and contribute to the psychophysiological synchronization of breeding partners.

COURTSHIP CHAIN: A series of courtship acts displayed in a fixed and typical sequence.

COURTSHIP GRUNTS (swine): Low frequency rhythmical sounds produced by mature boars when exposed to gilts or sows. Synonym: Chant-de-coeur.

COVERING: Mating in horses. (colloquial term)

COW: A bovine female after delivery of her first calf.

COYNESS: A ritualized short distance flight of females manifested as part of precopulatory displays and which generally stimulates a following response by males.

CRADLE: A wide collar, consisting of several strong wooden sticks with their ends interconnected by ropes so that the sticks are separated by intervals of space. A cradle is placed around an animal's neck to prevent it from licking its own wounds or removing bandages.

CRANIAL NERVES: Twelve pairs of nerves that emerge from the brain through various foramina of the skull and which have sensory or motor functions, or both.

CRANIAL PRESENTATION: Fetal presentation when only the nose enters the birth canal while both forelegs are retained below the body of the fetus.

CRATE: A housing compartment that permits lying down and standing up but prevents the animal from turning around or leaving the crate. As a housing system, crates presently are used mainly in swine production units during gestation (gestation crates), and during parturition and nursing to minimize overlying of piglets (farrowing crates). Crates are also used to house veal calves (veal crate), and for examination, weighing, and transportation of animals.

CREAMING: *See* Plant Avoidance. (colloquial term)

CREATIONISM: A theory which maintains that existing species were created as separate entities. Antonym: Evolutionism.

CRECHE: An aggregation of juvenile animals, typically birds, that have left their parental nests and band together.

CREEP FEEDING: Provision of feed supplements to young animals prior to weaning.

CREPUSCULAR: Pertaining to dawn and dusk. *Compare*: Diurnal, Nocturnal.

CREPUSCULAR ORGANISM: An organism which performs most of the behavioral actions in its repertoire around dawn and dusk and rests predominantly during the hours of daylight and darkness.

CRETINISM: Iodine deficiency of young animals causing exhaustion of the thyroid gland, inhibition of normal growth, and CNS malfunction. Behavioral symptoms are reduced alertness, hypoactivity and lethargy.

CRIBBING: Frequent biting on boards, pipes, and other structural materials used for confinement of animals, assumed to be indicative of boredom, pain, or nutritional deficiency. In horses, the term cribbing also refers to swallowing of air through the open mouth when biting on an object (also called crib biting or wind sucking).

CRIB BITING: *See* Cribbing.

CRITICAL PERIOD: An interval in the development of an organism during which it is most sensitive to particular experiences, (e.g., the interval during which an organism is optimally disposed to learn some task).

CRITICAL TEMPERATURE: A thermal threshold point below which an endothermic organism loses body heat to the environment at a greater rate than it is able to produce heat through metabolism (lower critical temperature), or above which an organism is unable to dissipate sufficient body heat to balance the heat generated by metabolism or gained from surroundings (upper critical temperature). In either case, the organism will not be able to maintain a stable body temperature and will die of hypothermia or hyperthermia, respectively.

CROP (poultry): An enlargement of the esophagus in birds which primarily serves as a reservoir for food and allows its moistening.

CROSSBREEDING: Sexual reproduction in which gametes originate from animals belonging to different breeds.

CROSSED REFLEX: A reflex in which stimulation of one side of the body causes an identical response on both sides of the body (frequently observed, for example, in eye reactions).

CROSS-FIRING (horse): Defective leg action occurring in pacers when the hoof of the hind-leg hits the fore-leg on the opposite side of the body.

CROSS PREGNANCY: A diagnostic term for development of a fetus in the horn of the uterus opposite to the side where ovulation occurred.

CROSS-SUCKING: Sucking performed by young, preweaned mammals but directed towards the mammary of lactating female other than their own dam.

CROUCH, SEXUAL: *See* Sexual Crouch.

CROUCHING: Lowering of the body by bending the legs.

CROUPADE (horse): A dressage exercise of the Spanish High Riding School performed as a vertical leap from the Pessage position. While off the ground, the horse folds its forelegs and draws up its hind legs before landing simultaneously on all fours at the same spot from which it leaped.

CROUP PRESENTATION: Fetal presentation when the posterior part of the body enters the birth canal while all legs are folded below the body of the fetus.

CROWDING: An unusually high spatial density of animals which may cause discomfort to some or all animals in the group, but not serious deprivation or injury. Reduced individual distance zones, for the most part, still can be maintained, and all animals are able to rest at the same time, stand up and lie down freely, extend their limbs without interference, and have adequate opportunity for eating and drinking. *Compare*: Overcrowding.

CROWDING DENSITY: *See* Spatial Density.

CROWING (chicken): A type of vocalization produced by males which has frequency oscillations of relatively wide amplitude, distinct breaks and a duration of approximately two seconds. It is assumed that crowing has a territorial and sexual function, and its frequency of occurrence bears a positive relation to social status in the flock. Crowing commences after six weeks of age.

CRURAL: Pertaining to the leg or thigh.

CRYPTIC BEHAVIOR: Any behavior that appears to be performed for the purpose of minimizing conspicuousness of an organism.

CRYPTIC COLORATION: Coloration of the body surface which makes an animal inconspicuous against the predominant background.

CRYPTORCHID: A male whose testes fail to descend into the scrotum.

CUD: A bolus of regurgitated feed to be remasticated and swallowed again.

CUDDING: Rumination. (colloquial term)

CUD-DROPPING: Tendency to drop regurgitated feed from the mouth.

Cud-dropping is one of the behavioral symptoms of pharyngeal paralysis.

CUE BLOCKING: In the context of conditioning, prevention of an association being made between a discriminative cue and a reinforcer by the concurrent presentation of another discriminative cue whose association to the reinforcer already has been firmly established.

CULTURAL TRANSMISSION: Non-genetic transfer of accumulated information between individuals. Through cultural transmission, adaptive knowledge and behavior patterns can become effectively fixed in a population without the mediation of genetic inheritance.

CURIOSITY: A tendency to approach and investigate novel stimuli or situations and become familiar with their attributes or implications.

CURLED TONGUE (turkey): Deformity of the tongue in young poults that causes feeding difficulties.

CURSORIAL: Adapted for running.

CURVE: *See* specific curves—DARK ADAPTATION; LEARNING; NORMAL; OGIVE.

CYCLE: *See* specific cycles—AHEMERAL; BEHAVIORAL; ESTRUS; LD; NURSING; RESPIRATION.

CYCLICAL NURSING: Nursing conducted at a regular time interval.

CYSTIC OVARIAN DISEASE: A reproductive disease of cattle characterized by the presence of cystic structures in the ovaries resulting in infertility. Behavioral symptoms include nymphomania, restlessness and aggressiveness.

D

DAM: The female parent.

DANCE: *See* specific dances—BEE; GROOMING; SHAKING; SICKLE; WAGGLE.

DANCE LANGUAGE: A general term for any repeated series of rhythmical movement or locomotion which is performed as a means of communication.

DARK ADAPTATION: A process whereby the eye becomes increasingly sensitive to visual stimulation in poorly illuminated visual fields.

DARK ADAPTATION CURVE: The curve defined by the change in minimum light intensity perceivable by an organism over increasing time spent in darkness. In species having both cones and rods in the retina of the eye, the dark adaptation curve characteristically undergoes a sudden change in slope when the shift from photopic to scotopic vision occurs.

DARWINISM: A theory of biological evolution emphasizing the role of natural selection.

DAYS OPEN: A term designating the length of time from parturition to new conception.

DEAFFERENTATION: Destruction of the ability of afferent nerves to transmit action potentials. Deafferentation techniques are used to study the function of nerve receptors and afferent nerves and their impact on the central nervous system and behavior.

DEAFNESS: Undeveloped or severely inhibited ability to hear.

DEATH FEIGNING: *See* Lethisimulation.

DEBEAKING (poultry): Beak trimming. (colloquial term)

DEBILITY: Severe loss of or complete lack of strength.

DECISION-MAKING PROCESS: A cognitive process of integration of sensory inputs and available memory information for the purpose of determining a course of action.

DECLARATIVE REPRESENTATION: A hypothesis that objects, goals, or relationships in the environment are represented in the mind as propositions of the reality (object hypotheses). Declarative representation does not accentuate how such propositions should be used. *Compare:* Procedural Representation.

DECLAWING: Surgical removal of claws from the feet of birds, cats, and dogs, to prevent scratching and injuring others.

DECORTICATION: Surgical removal of the cerebral cortex.

DEDUCTION: A conclusion reached through reasoning.

DEEP BEDDING: *See* Deep Litter.

DEEP LITTER: Microbiologically active bedding material consisting of porous substances (e.g., wood shavings, cut straw, etc.) able to absorb moisture from excreta of animals and maintain a relatively constant temperature over a long period of time. Deep litter is often used in loose housing systems.

DEFECATION: Elimination of feces from the body. During defecation, healthy animals raise the tail head and may also move the tail slightly to one side. Horses, cattle, sheep, and goats may defecate in a stationary position or while walking. Swine normally remain stationary and tend to deposit feces in specific areas of a pen, such as in a corner, in a place where feces already have been deposited or in wet areas.

DEFECT: An absence, failure, or imperfection in comparison with some standard.

DEFENSIVE AGGRESSION: Aggression by a victim or potential victim directed toward an instigator or perceived instigator of an aggressive interaction. Defensive aggression characteristically is manifested to prevent or reduce some annoyance or loss.

DEFENSIVE BEHAVIOR: Behavior performed to prevent or neutralize a real or perceived aversive stimulus. According to the circumstances, such behavior may encompass aggression, avoidance, or signs of appeasement and subordination.

DEFERRED REACTION EXPERIMENT: An experimental strategy designed to assess the ability of an organism to memorize cues temporarily linked with a familiar stimulus (e.g., a subject exposed to several identical objects may be tested to determine the length of time it remains able to identify the one which was briefly illuminated). Synonym: Delayed Response Method.

DEFLATION REFLEX: A respiratory reflex controlling expiration of air from the respiratory tract.

DEFLEECING: Removal of the fleece from a sheep's integument. This term is used most frequently to refer to removal of fleece by chemical means.

DEFOLIATION: A form of selective grazing whereby animals consume only the leaves of plants.

DEFOLIATION, PROGRESSIVE: *See* Progressive Defoliation.

DEGENERATION: Gradual deterioration in the structure or efficiency of the organs, the nervous system, or the whole body of an organism.

DEGLUTITION: Swallowing.

DEHORNING: Surgical removal of horns or any treatment which prevents the development of horns in farm animals.

DELAYED CONDITIONING: Classical conditioning wherein the presentation of the conditioned stimulus (CS) begins considerably before the unconditioned stimulus (UCS). The conditioned response occurs as a consequence of the CS, but also may be timed, apparently cued by an internal clock, relative to the expected presentation of the UCS.

DELAYED RESPONSE METHOD: *See* Deferred Reaction Experiment.

DELIBERATION: The initial stage of the choice-making process.

DELIVERY: *See* Parturition.

DELIVERY, CAESAREAN: *See* Caesarean Delivery.

DELIVERY EASE: A technical term for subjective assessment of the parturition process on a ranking scale from easy to difficult delivery. Delivery ease is considered an important reproduction variable, particularly in cattle, and if expressed numerically, can be averaged per individual animal (dam or sire), herd, breed, or season.

DELTA ACTIVITY: *See* Brain Wave Activity.

DEMAND FUNCTION: A function describing the amount of a commodity an animal will work for relative to the cost of the commodity. If the amount of the commodity acquired is independent of cost, the demand for the commodity is inelastic. If the converse is true, the demand is elastic. Actions involved in the acquisition of commodities for which there is inelastic demand tend to have high behavioral resilience, and vice versa.

DENDRITE: A fibrous extension of a neuron able to receive and transmit action potentials toward the main body of the neuron.

DENSITY: *See* specific density—POPULATION; SPATIAL.

DENSITY DEPENDENCE: A phenomenon in which the expression of a behavior is affected by the amount of space available per animal, e.g., frequency of wing flapping by chickens in cages. *Compare:* Group Size Dependence.

DENTITION: A term used in reference to conformation and configuration of the teeth.

DEONTOLOGICAL MORAL THEORIES: Theories that claim that human actions are morally right or wrong if they respect or violate someone's inherently valuable interests, or respect or violate divine command. The animal rights and divine command theories are examples of deontological moral theories. *Compare:* Consequentionalist Moral Theories.

DEPARTURE, MATERNAL: *See* Maternal Departure.

DEPENDENCE: *See* specific dependence—DENSITY; DISTANCE; ENVIRONMENTAL COMPLEXITY; GROUP SIZE; SOCIAL.

DEPENDENT VARIABLE: Any variable which changes as a result of the effects of an independent variable(s). In a behavioral study the dependent variable is a measurable response.

DEPRAVED APPETITE: *See* Pica.

DEPRESSION: In a behavioral sense, a state of severe emotional dejection characterized by various behavioral disorders. *Also see* specific depressions—AGITATED; ENZOOTIC; STUPOROUS.

DEPRIVATION: Removal of needed substances (feed deprivation, water deprivation), perceptual isolation from things desired (social isolation), or prevention of the performance of necessary behaviors (sleep deprivation, exercise deprivation). Deprivation frequently is used experimentally to induce some detectable drive.

DESENSITIZATION: Decrease of a response to a given stimulus with practice or number of trials. Antonym: Sensitization.

DESIRE: Motivation of an organism to acquire, control, or experience some feature of its environment (e.g., water, feed, space, social contact, escape from aversive stimuli, etc.). Desires are interpreted operationally from observed behavior and may, in turn, be indications of underlying needs. *Compare:* Need.

DESNOODING (turkey): Surgical removal of the snood in neonatal poults. Such a treatment is performed to reduce subsequent incidence of cannibalism.

DESPOT: A dominant individual greatly restricting the activities of subordinates. In an extreme theoretical case, the social hierarchy would consist of one alpha animal and subordinates having no rank differences among themselves.

DETECTION, PREGNANCY: *See* Pregnancy Detection.

DETECTION THRESHOLD: The weakest intensity of a stimulus that can be sensed. The term detection threshold has been substituted for absolute threshold because the threshold proves to be probabilistic rather than absolute. *Compare:* Threshold.

DETECTOR, HEAT: *See* Heat Detector.

DETERMINISTIC: Referring to a fixed relationship between two or more variables. It does not admit the possibility of chance affecting this relationship. Antonym: Stochastic.

DETRACTION: In a strict behavioral sense, lowering attention or concentration without shifting the point of focus.

DEVIATION: Variation from a point of reference, norm, or standard.

DEWATTLING (poultry): Surgical removal of a fowl's wattles.

DEXTRAL: Pertaining to the right side of the body. Antonym: Sinistral.

DIAGNOSIS: Identification of a disease or behavioral disorder.

DIAGONAL GAIT: A gait in which the diagonal legs move at the same time (e.g., trot).

DIARRHEA: Excessively fast flow of alimentary content through the digestive tract manifested by frequent defecation of highly liquid excreta. Diarrhea is generally a serious symptom of illness and if not cured leads to dehydration and death of the affected animal.

DIASTOLE: The period of dilation of the heart associated with the filling of the ventricular chambers with blood.

DICHO-: Prefix meaning "double" or "segregation into two."

DICHOTOMOUS DISTRIBUTION: Classification of observations into two mutually exclusive categories.

DIESTRUS: A relatively short period of sexual inactivity between two estrus cycles in polyestrous animals.

DIET FRACTIONATION: In a behavioral context, preferential consumption of certain components of a diet on the basis of nutritional content, palatability, particle size, shape, color, smell, etc. Diet fractionation may lead to unbalanced nutrition if feed supply is overabundant, or if feeder space is limited so that all animals cannot eat at the same time.

DIFFERENCE: *See* specific difference—JUST NOTICEABLE; MORALLY RELEVANT.

DIFFERENCE THRESHOLD: The minimum change in intensity of stimulation required to cause a noticeable change in sensation.

DIFFERENTIAL REINFORCEMENT: The procedure used in operant conditioning to establish a discrimination between two stimuli. For example, a pigeon can be trained to discriminate between red and green discs if it is reinforced for pecking one but not the other.

DIGESTION: The process involving both chemical and physical action by which food is broken down in the digestive tract into smaller molecular components that can be absorbed into the circulatory system.

DIGGING: *See* Excavation.

DIGITIGRADE LOCOMOTION: Locomotion conducted on the tips of the toes. *Compare:* Plantigrade Locomotion.

DILATATION STAGE (Labor): *See* Labor.

DIMINISHED APPETITE: *See* Reduced Appetite.

DIPLOID: Containing two sets of homologous chromosomes. Somatic cells are diploid cells. *Compare:* Haploid.

DIPLOPIA: Double vision of a single object.

DIPPING: Brief, complete or partial immersion of animals into a disinfecting fluid to prevent or control infestation of ectoparasites. *Also see* specific dipping—HEAD; TEAT.

DIRECTED SIGNAL: A signal that is emitted to convey information to a specific target individual(s) (e.g., threat signal). *Compare:* Broadcast Signal.

DIRECTIONAL SELECTION: Selection favoring organisms with characteristics that are to one side of the mean of the population frequency distribution for such characteristics. The effect is to move the mean of the frequency distribution toward the favored phenotype representing the highest genetic fitness or best production performance.

DIRECT MEASUREMENT: The measurement of a variable without transformation of the resultant datum to another scale prior to analysis; the datum resulting from such measurement.

DIRECT OBSERVATION: Attention to ongoing phenomena, for the purpose of recording behavior, without aid of intermediary technology (e.g., *in situ* visual observation of an organism's movement, auditory observation of its vocalizations, etc.).

DIRECT REFLEX: A type of reflex in which the receptor and effector are located on the same side of the body.

DISCRETE: Separate or discontinuous.

DISCRETE DISTRIBUTION: A frequency distribution consisting of indivisible units (e.g., number of responding subjects, number of positive responses).

DISCRETE SIGNAL: A signal whose information is determined only by the fact that the signal is emitted, i.e., the information content is not altered by some form of signal modification. *Compare:* Graded Signal.

DISCRIMINAL PROCESS: The process of discrimination between stimuli.

DISCRIMINATION, SENSORY: *See* Sensory Discrimination.

DISCRIMINATION TIME: The time elapsing between exposure of a subject to a discriminatory situation and the instigation of a discriminatory response.

DISCRIMINATION TRAINING: A type of training where the subject is reinforced (through reward or punishment) to discriminate between two or more stimuli.

DISCRIMINATIVE CUE: A stimulus that occasions an appropriate response as a consequence of conditioning. For example, if a specific instrumental response is rewarded only when attendant wears certain clothes, the clothes may become a discriminative cue.

DISCRIMINATIVE PUNISHMENT TRAINING: *See* Discriminative Training.

DISCRIMINATIVE REWARD TRAINING: *See* Discriminative Training.

DISCRIMINATIVE STIMULUS: *See* Discriminative Cue.

DISCRIMINATIVE TRAINING: Operant conditioning in which application of response-contingent reinforcement is linked to another stimulus (discriminative cue) such that responses are reinforced in the presence (or absence) of the discriminative cue. When reinforcement is positive, the process is called discriminative reward training, when negative, it is called discriminative punishment training.

DISEASE: A state of pathologically abnormal physiological and/or psychological functioning. *Also see* specific diseases—ACUTE; CHRONIC; ENZOOTIC; FOOT-AND-MOUTH; GREASY PIG; MAD COW; NEWCASTLE; PRODUCTION; PSYCHOSOMATIC.

DISHABITUATION: Recovery of responsiveness during the course of habituation when a novel, previously irrelevant stimulus appears at the same time as the stimulus to which habituation is occurring.

DISHING (horse): Defective leg action when the leg is thrown inward during the extension phase.

DISHORNING: *See* Dehorning.

DISINHIBITION: Reoccurrence of suppressed response to the original eliciting stimulus due to motivational interference of another, previously irrelevant stimulus.

DISINHIBITION, PAVLOVIAN: *See* Pavlovian Disinhibition.

DISORDER, *See* specific disorders—BEHAVIORAL; FUNCTIONAL; ORGANIC.

DISORIENTATION: Temporary or permanent inability of an organism to perceive the spatial arrangement of its own surroundings and to maintain relevant point(s) of reference.

DISPERSION: Diminishing spatial proximity among organisms. Also, the particular positional arrangement of such organisms (e.g., tight dispersion, loose dispersion, etc).

DISPLACEMENT: Behavior that is seemingly irrelevant to the situation being faced by an animal. It may occur when an animal is frustrated in its attempts to perform actions appropriate to the circumstances, or when it apparently experiences conflict among motivations for different actions. It is also postulated that displacement involves motivational disinhibition.

DISPLAY: *See* specific display—BEHAVIORAL; CONFLICT THEORY OF; DIVERSIONARY.

DISPOSITION: All the genetically influenced tendencies and propensities of an individual or a group. *Also see* specific disposition—GENETIC; LEARNING.

DISRUPTIVE SELECTION: Selection favoring more than one phenotype, when such phenotypes do not represent the majority of the population, with intermediates being selected against. The effect is to increase the phenotypic variance in the population.

DISTAL: Remote or farther from a point of reference. In an anatomical sense, the point of reference might be the central part of an animal's body. Antonym: Proximal.

DISTAL STIMULUS: Any stimulus which acts on sensory receptors through a proximal stimulus (e.g., a tree in a visual field is a distal stimulus, the image of this tree on the retina is a proximal stimulus). *Compare:* Proximal Stimulus.

DISTANCE: *See* specific distance—FLIGHT; INDIVIDUAL; SOCIAL..

DISTANCE DEPENDENCE: A phenomenon in which the expression of a behavior is determined by the distance of an individual or group to other organisms or objects. *Also see* specific distance—SOCIAL; FLIGHT.

DISTANT SPECIES: Species whose members prefer a more solitary lifestyle or less social involvement than is observed in contact species.

DISTOCEPTOR: Receptors which respond to distal stimuli (e.g., visual, audio, thermal receptors).

DISTORTION: Any misinterpretation of reality. Perceptual distortions are commonly called illusions.

DISTRACTION: A situation where concentration of attention is interrupted or disturbed by some stimulus.

DISTRESS: An emotional state of an organism resulting from excessive fear, loss of companion or object with which it has a strong psychological bond, physical discomfort, food or water deprivation, pain etc. *Compare:* Suffering.

DISTRESS CALL: A vocalization indicative of the experience of distress on the part of the vocalizing animal. A distress call may or may not be a signal.

DISTRESS SIGNAL: A signal indicative of the experience of distress on the part of the organism emitting the signal.

DISTRIBUTION: Systematic arrangement of numerical observations according to some ranked measurement scale. The term is also used in the description of the spatial arrangement of events or organisms. *Also see* specific distribution—CONTINUOUS; DISCRETE.

DISTURBED BEHAVIOR: *See* Abberant Behavior.

DISUNITED CANTER (horse): A three-beat gait which differs from the normal canter in the second beat when the movement of the legs is synchronized laterally rather than diagonally. *Compare:* Canter.

DIURESIS: Increased formation of urine.

DIURETIC: Any agent which increases formation of urine.

DIURNAL: Pertaining to the daylight hours. *Compare:* Crepuscular, Nocturnal.

DIURNAL ORGANISM: An organism that performs most of the behavioral actions in its repertoire during the daylight hours and rests predominantly at night.

DIVERGENT SELECTION: Isolation of two sub-populations from some founding population and selection within each such that the phenotype favored in one sub-population is the opposite alternative to that favored in the other. Divergent selection sometimes is used to assess the effectiveness of a selection program. Synonym: Bidirectional Selection.

DIVERSIONARY DISPLAY: An activity which serves to distract a potential predator from vulnerable prey, e.g., offspring.

DIVINE COMMAND THEORY: A theory guided by the principle that an action can be identified as morally right or wrong as a result of being prohibited, required, or permitted by divine authority. According to this theory guidelines for the treatment and use of animals can be obtained, for example, from interpretation of Holy Scripture.

DOCILE: Obedient or submissive.

DOCKING: Partial or complete amputation of the tail.

DOCTRINE, CARTESIAN: *See* Cartesian Doctrine.

DOE: A female goat, deer, or rabbit after delivery of its first offspring.

DOGMA: An uncontested belief accepted to be true because it was proclaimed to be true by an authority or strong tradition. For many people the belief that humans have dominion over animals is a dogma.

DOG-SITTING POSITION: Resting on the caudal part of the body observed in pigs, cattle, and horses. It is often, but not always, indicative of disease (e.g., partial paralysis due to a vitamin or mineral deficiency, abdominal pain, etc.).

DOG-SITTING PRESENTATION: Fetal presentation when all four limbs and the nose enter the birth canal concurrently.

DOMESTICATED ANIMAL: Any animal whose progenitors have undergone a domestication process, and which itself is kept under direct human control. *Compare:* Feral Animal.

DOMESTICATION PROCESS: An evolutionary process during which the living space, animal care, selection of mating partners, and reproduction of a species of animal has become supervised or controlled by humans. As the process continues, the species is genetically altered from its ancestral form.

DOMESTICATION TRAIT: Any genetically controlled qualitative or quantitative trait of an organism that has developed as a consequence of the domestication process.

DOMICILE: A larger area, extending on all sides of a territory, that serves as a cruising or ambulatory range, but is not defended.

DOMINANCE: *See* Social Dominance.

D

DOMINANCE AGGRESSION: Aggression manifested in the context of establishment, maintenance, or improvement of social status.

DOMINANCE BOUNDARY: A theoretical concept referring to the point at which the ratio between the strengths of motivations for two alternative actions is such that each has an equal tendency to be the dominant activity. The ratio rarely would remain on the dominance boundary; normally one activity would become dominant for a while, and the other subdominant, with time-sharing being manifested accordingly.

DOMINANCE SIGNAL: Any behavioral display or sign indicative of seeking or maintaining dominance over another organism(s). Common dominance signals are upright posture, intense staring, exposure of weapons, and many of the broad variety of threat signals.

DOMINANCE ORDER: *See* Social Hierarchy.

DOMINANT ACTIVITY: A behavioral action which in motivational competition superimposes alternative action. The dominant action controls the expression of these alternative actions through inhibition or disinhibition.

DOMINANT HEMISPHERE: The hemisphere of the brain which exercises the most control over a given sensation or behavior.

DOPPLER APPARATUS: An apparatus designed to detect fetal movement using ultrasound.

DORSAL: Pertaining to, or in the direction of, the dorsum (back) of an organism.

DORSAL TRANSVERSE PRESENTATION: Fetal presentation when the dorsal part of the fetus approaches the birth canal.

DORSIFLEXION: Bending upward toward the dorsum (or forward in leg motion).

DOUBLE MUSCLING: A genetically controlled greater than usual number of fibers in skeletal muscles and consequent enlarged amount of muscle tissue, particularly in the hind quarters. Double muscling is more common in some breeds of cattle than others (e.g., Belgian Blue, Charolais, Piedmont) and may cause extended gestation period and dystocia.

DOUBLE PREGNANCY: *See* Superfetation.

DRAFT ANIMAL: An animal used as a source of kinetic power (pulling loads, plowing, etc.).

DRAIZE TEST: A test to determine a body's reaction to a substance by applying it to one of twin organs while the other serves as a control, e.g., placement of a substance in an eye of a rabbit to assess the degree of irritation that develops by comparison with the condition of the other eye. The Draize test has been criticized as being unnecessarily harmful to the well-being of test animals because the reasons for the testing of many substances have not been sufficiently weighty and because the test results may not be generalizable to the human biological system to which they are intended to apply.

DRAKE: A sexually mature male duck.

DREAM: A series of hallucinatory experiences.

DRENCHING: Oral application of liquid medicines to animals. (colloquial term)

DRESSAGE (horse): A set of standardized exercises developed through a training program which stresses locomotory coordination of the horse, as well as harmony between horse and rider. *Also see* Piaffe; Pirouette; Travers.

DRIFT, BEHAVIORAL: *See* Behavioral Drift.

DRIFT, SOCIAL: *See* Social Drift.

DRINKING: Voluntary oral ingestion of liquids.

DRIVE: To force one or more animals to move from one place to another and to direct their progress; also impetus for behavior generated within an organism as a consequence of neural activation due to internal or external stimulation. Such stimulation may arise from physiological disbalance (caused e.g., by food or water deprivation) or other internal or external events. Drive refers to the initial component of an organism's hypothetical behavioral response mechanism. *Compare:* Motivation. *Also see* specific drive—PRIMARY; SECONDARY.

DRONE: A male bee.

DROP, LEG: *See* Leg Drop.

DROPPED ELBOW: Radial paralysis. (colloquial term)

DROWSING: Being in a somnolent state characterized by reduced attention, eye closure, muscular relaxation, and immobility.

DRUMMING: A rapid series of repeated nonvocal sounds. Geese may produce such sounds by vigorous flapping of the wings.

DRY: A nonlactating mature female. This term most frequently refers to nonlactating cows prior to their next parturition.

DRY COAT: *See* Anhydrosis. (colloquial term)

DUAL IMPRESSION: Reception of stimuli originating from the same source by two types of sensors.

DUBBING (poultry): Trimming of the comb. Such a treatment is performed, generally on neonatal birds, to reduce subsequent incidence of cannibalism and comb injuries.

DUCK: A mature female duck.

DUCKLING: A young duck.

DUE DAY: The expected day of parturition. Due day is calculated using the average duration of pregnancy of a given species or breed.

DUMMY SYNDROME: A clinical term occasionally used to refer to decreased or diminished awareness of an obstruction in the path of movement.

DUST-BATHING (chicken): A behavior pattern that is a component of integumentary care characterized by lying on the side and making a small depression in the floor surface while head rubbing, bill raking, wing shaking, and scratching on the floor. During dust bathing, birds throw particles of floor material over their bodies and the surrounding area. The dust bathing event ends with ruffling and preening. Synonym: Sand bathing, Litter bathing. *Also see* specific dust bathing—SHAM.

DYAD: Two organisms reacting as a single unit due to a strong interdependence (e.g., a pair of pulling horses).

DYNAMIC PSYCHOLOGY: A school of psychology emphasizing the central role of motives and drives in the behavior of organisms.

DYNAMICS, See specific dynamics—BEHAVIORAL; POPULATION; SOCIAL.

DYSKINESIA: Inability to perform a voluntary movement in its usual complete form.

DYSMETRIA: Serious difficulty or inability to assess distances correctly and execute movements in their proper spatial context. *Compare:* Hypermetria, Hypometria.

DYSPHAGIA: Serious difficulty or complete inability to swallow.

DYSPHONIA: Serious distortion or complete dysfunction of vocalization.

DYSPNEA: Difficult breathing.

DYSTOCIA: Difficult parturition which may be caused by mechanical obstruction of the birth canal (cervical dystocia), abnormal shape or position of the fetus (fetal dystocia), or difficulty in expulsion of the placenta (placental dystocia).

DYSTROPHY: Progressive atrophy of tissues.

E

EAR: A sensory organ for hearing and postural balance consisting of the external ear (auricle and meatus), middle ear (tympanic membrane, tympanic cavity, and tympanic ossicles), and internal ear (vestibule, cochlea and semicircular canals).

EAR BITING: Biting or chewing on the ears of other animals. It occurs most frequently in early-weaned, trough- or bucket-fed young mammals, particularly calves and piglets. Ear biting can be substantially reduced if the young animals are fed through artificial nipples. Ear biting may lead to serious injury of the victim's ear(s).

EAR DRUM: *See* Tympanic Membrane.

EAR FLAPPING: Rapid rotatory oscillation of the head causing the ears to flap and strike against the face. Ear flapping is relatively common in swine, and it is conjectured to be a display of mild excitement, a threat signal, or a response to irritation of the ear caused by parasites or specific sounds.

EAR FLICKING: Quick, light movements of one or both ears.

EARLY WEANING: Artificially induced permanent separation from the dam or surrogate prior to the time of natural weaning.

EARS, RETRACTED (horse): *See* Retracted Ears.

EAR SUCKING: Sucking of the ears of littermates or conspecifics. Ear sucking occurs frequently in early-weaned, trough- or bucket-fed piglets or calves, or piglets of hypogalactic dams. It may cause lesions of the ear and in crowded pens may lead to serious abscesses.

EATING: Voluntary oral ingestion of solids and semi-solids.

EATING, EGG: *See* Egg Eating.

ECG (EKG): *See* Electrocardiogram.

ECHO-: A prefix used to express repetition or imitation (e.g., echopraxia).

ECHOPRAXIA: Imitation of the movement of another organism.

ECOLOGY: The study of the relationships between organisms and their environment. Synonym: Bionomics.

ECOSYSTEM: The complete array of interacting biological and nonbiological factors occurring in a given geographical area.

EEG: *See* Electroencephalogram.

EFFECT: *See* specific effect—BRUCE; COOLIDGE; FRASER DARLING; HALO; HETEROSIS; LEE-BOOT; PROXIMITY; VANDENBERG; WHITTEN.

EFFECTIVE ENVIRONMENTAL TEMPERATURE: The cumulative effect of all aspects of an environment on the heat balance of an animal determined by comparison to, and expressed as, the temperature of an isothermal, still environment that has the same thermal effect on the animal. Effective environmental temperature frequently is not the same as the air temperature of the environment because of modifying influences, e.g., thermoregulatory behavior of the animal, temperature of radiant surfaces in the environment, air movement, and relative humidity. The responses of animals (e.g., alteration of exposed body surface, panting, etc.) to the thermal aspect of a given environment are on the basis of the effective environmental temperature.

EFFECTOR: The organ (muscle, gland) which carries out a function in response to efferent action potentials.

EFFERENCE COPY: A hypothetical set of expected consequences of behavior derived both from past experience and current perceptions of the environment against which an animal presumably compares the actual consequences of its behavior.

EFFERENT: Carrying outward; concerned with the transmission of action potentials from the central nervous system to the reacting mechanisms (muscles and glands).

EFFERENT NERVES: Nerves that transmit action potentials toward muscles and glands. Also called motor nerves. *Compare:* Afferent Nerves.

EGG BEATER (horse): A horse having very light and coordinated leg action. (colloquial term)

EGG EATING (poultry): A vice characterized by penetration of the egg shell by pecking and partial consumption of the egg content.

EGG LAYING (poultry): *See* Oviposition.

EGG ROLLING (poultry): Retrieval of eggs to the nest or periodic movement of incubating eggs in the nest by breeding birds.

EGOISM: A theory that the behavior of organisms is guided by self-interest (psychological egoism). In an ethical sense this term implies that human action can be assessed as morally right or wrong according to whether it benefits or harms the person who acts (ethical egoism).

EGOISM, RATIONAL: *See* Rational Egoism.

EJACULATION: A sudden discharge, commonly referring to the expulsion of semen during male orgasm. *Also see* specific ejaculation—PREMATURE.

EJACULATIO PRAECOX: *See* Premature Ejaculation.

EJACULUM: Semen discharged during one ejaculation.

EJECTION: The process of discharging or being discharged (e.g., milk ejection). *Also see* specific ejection—MILK.

EKG (ECG): *See* Electrocardiogram.

ELBOW, DROPPED: *See* Dropped Elbow.

ELECTROCARDIOGRAM: A record of the electrical activity generated by the heart muscle.

ELECTROEJACULATION: Any ejaculation artificially induced by electrical stimulation. Electroejaculation is occasionally used for the purpose of semen collection.

ELECTROENCEPHALOGRAM (EEG): A record of the electrical activity generated by the nerve cells of the brain.

ELEMENTARISM: An attempt to explain behavior from the position of elementary psychological processes. This term is also used to describe excessive preoccupation with simple elements at the expense of the whole.

ELIMINATIVE BEHAVIOR: Behavior involved in the expulsion of feces or urine from the body or dwelling place.

ELONGATED ESTROUS CYCLE: An estrus cycle extended beyond its typical range for a given species.

EMBRYO: A prenatal organism in the phase during which the major body structures are being formed. *Compare:* Fetus.

EMBRYONIC BEHAVIOR: *See* Prenatal Behavior.

EMBRYONIC MALPOSITION (poultry): Deviant fetal position in hatching eggs leading to mortality or poor viability. These malpositions generally are classified into six groups: I—head positioned over right wing, II—head turned to the left, III—head directed away from the air cell, IV—one leg over the head or beak, V—head in the small end of the egg, and VI—head between the thighs. *Compare:* Normal Embryonic Position.

EMBRYONIC POSITION, NORMAL: *See* Normal Embryonic Position.

EMBRYO TRANSFER: Placement of an embryo taken from a donor female into the uterus of a suitably prepared recipient female for gestation and delivery. Such transfer is done to propagate superior genotypes and also allows investigation of intrauterine influences on behavior and production performance.

EMESIS: An act of vomiting.

EMETIC: Pertaining to or inducing vomiting.

EMETIC REFLEX: The vomiting reflex of monogastric animals which results from aversive stimulation of the digestive tract.

EMIGRATION THRESHOLD: A theoretical level of arousal potential necessary for a lasting departure of an organism from its social unit.

EMOTION: A state of arousal, referred to as fear, anger, pleasure, joy, etc., associated with specific cognitive processes and characterized by specific behavioral and physiological symptoms.

EMOTIONAL BEHAVIOR: Any behavior that expresses a strong emotion. An animal may become incoherent in its ability to cope with the situation if it experiences a strong emotion.

EMOTIVISM: A theory that ethical statements or judgements are merely expressions of emotions. Proponents of this view claim that it is a theory concerning meaning of ethical terms (metaethical theory) rather than theory about what is right or what behavior is morally required.

EMPATHY: The ability to understand or assess the feelings of others.

EMPIRICAL: Based on experience rather than on theoretical considerations.

EMPIRICISM: A theory that considers experience as the sole basis of knowledge.

EMULATION: An attempt to equal or excel the performance of other organisms.

ENCEPHALITIS: Inflammation of the brain. Behavioral symptoms usually start with abnormal excitability followed by periods of depression, then recumbency and paralysis.

ENCEPHALOMYELITIS: Inflammation of the brain and spinal cord. Behavioral symptoms may include progressive stiffness, staggering locomotion, paralysis, and ultimately, death.

ENDOCRINOLOGY: The study of hormones and their role in the physiology and behavior of organisms.

ENDOGENOUS RHYTHM: A biorhythm activated by events originating within an organism. An endogenous rhythm does not require periodic external input to maintain its oscillation.

ENDOGENOUS STIMULATION: Stimulation arising from events occurring inside an organism's body.

ENDOTHERM: An organism that derives its body heat primarily from endogenous metabolic activity.

END-PLATE: A complex of branching nerve terminals that innervates a muscle fiber.

ENEMA: Infusion of fluid into the rectum and colon to relieve constipation.

ENERGY, ACTION-SPECIFIC: *See* Action-Specific Energy.

ENGINEERING, GENETIC: *See* Genetic Engineering.

ENRICHMENT, ENVIRONMENTAL: *See* Environmental Enrichment.

ENTERITIS: Inflammation of the intestines leading to diarrhea.

ENTOPTIC: Referring to processes occurring within the eye.

ENTRAINMENT: The process of development or the maintenance of a close sequential link between the temporal occurrence of two or more events.

ENURESIS: Involuntary urination during extreme emotional experiences or during sleep.

ENVIRONMENT: The total sum of nongenetic factors which interact with the genotype of an organism. *Also see* specific environment—BARREN; EXTERNAL; INTERNAL.

ENVIRONMENTAL COMPLEXITY: The diversity and intensity of environmental stimuli relevant to a given organism, age group, species, etc. Environmental complexity may range from very low to very high, and thus be characterized as insufficient, adequate, or excessive.

ENVIRONMENTAL COMPLEXITY DEPENDENCE: A phenomenon in which the expression of a behavior is affected by environmental complexity, e.g., manifestation of stereotypy, response to novel stimuli.

ENVIRONMENTAL ENRICHMENT: The process or factor which increases the complexity of the environment.

ENZOOTIC DEPRESSION: A clinical form of psychological depression that characteristically occurs in association with certain management systems or localities.

ENZOOTIC DISEASE: A disease of animals that characteristically occurs in certain geographic areas or management systems.

ENZOOTIC PNEUMONIA: A common form of pneumonia in young animals (especially pigs and calves), usually transmitted from the dam to her offspring, or between young animals in group housing. Symptoms include coughing, nasal discharge, fever, elevated respiratory rate, and in extreme cases, retarded growth.

EPIDEMIC: Spreading rapidly from one organism to another and, as a result, affecting a large number of organisms simultaneously.

EPIDEMIC TREMOR: Avian encephalomyelitis. (colloquial term)

EPIGAMIC: Pertaining to characteristics or stimuli which attract the opposite sex (e.g., body coloring).

EPIGENESIS: A concept of ontogeny in which phenotypic development is postulated to proceed through stages. Development within each stage is dependent upon three attributes: the genotype, the phenotype at the beginning of a given stage, and the environment during the stage.

EPIMELETIC BEHAVIOR: Behavioral activities associated with the provision of attention, care, or help to other individuals. Synonym: Care-giving Behavior.

EPIMELETIC BOND: *See* Bond.

EPINEPHRINE (NOREPINEPHRINE): Hormones of the adrenal medulla which stimulate physiological effects similar to those caused by the sympathetic nervous system and which mobilize bodily reserves in emergencies. Norepinephrine also helps to maintain the tone of the vascular system and plays a role in maintenance of blood pressure. Synonym: Adrenalin and Noradrenalin.

EPIPHENOMENALISM: A theory about the mind which implies the mental processes are effects of physical processes, but that mental processes do not have any effect on physical processes.

EPIPHORA: An overflow of tears down the cheek, often due to blockage of the tear ducts that normally direct the flow of tears from the eye down into the nasal cavity.

EPISTAXIS: Nose bleeding. Nose bleeding can be a symptom of lesions in the nasal cavity, auditory tube, nasopharynx, or lungs.

EPIZOOTIC: Pertaining to infectious diseases which may spread rapidly and affect large numbers of animals.

EPURATION: Emptying, clearing, or removing some internal substance. In a behavioral sense this term is used to refer to emitting a jet of bodily liquids, such as urine, at another organism. Epuration may be a defense strategy or a component of sexual behavior.

EQUAL CONSIDERATION OF INTERESTS: A moral principle based on the assumption that creatures have interests or preferences that have to be taken into consideration when determining human obligations toward them. The principle of equal consideration of interests is used by utilitarians to show that favoritism or discretion continent on features such as sex, race, species, etc. is morally unacceptable.

EQUESTRIAN: A term generally pertaining to the use of horses, particularly, horse riding.

EQUIFINALITY: Refers to a phenomenon in which some aspect of the phenotype, such as a particular behavior pattern or morphological attribute, is able to develop via more than one pathway.

EQUILIBRIUM: A state of balance in which opposing forces or tendencies counteract each other.

EQUINE INFLUENZA: A highly infectious viral disease of horses. Symptoms include coughing, fever, depression, and occasionally, muscular pain.

EQUIVOCAL: A statement or occurrence that may have two or more equal meanings. In a behavioral context this term is used in motivational assessment to refer to any behavioral action that has two or more different but equally probable explanations or functions.

ERECTION: A process or state of becoming rigid or elevated. This term commonly refers to swelling of erectile tissues such as the clitoris, nipples, and penis.

ERGOGRAM: A record of muscular exertion.

ERGONOMICS: The study of the division of labor within a species, e.g., as observed in bees, ants, etc.

EROGENOUS ZONE: *See* Erotogenic Zone.

EROTIC: Referring to sexual stimuli, sexual sensations, or sexual excitement.

EROTOGENIC ZONE: A body area on an organism that when appropriately stimulated results in the organism becoming sexually aroused. Synonym: Erogenous Zone.

ERRONEOUS IMPRINTING: Imprinting on objects other than those natural for a given organism. Erroneous imprinting seldom occurs in conditions of normal maternal care in a natural habitat.

ERROR, INSTRUMENTAL: *See* Instrumental Error.

ERUCTATION REFLEX: The casting up of gases from the stomach through the mouth, induced by internal pressure.

ESCAPE: Speedy departure from a place of aversive or potentially aversive stimulation.

ESOPHAGITIS: Inflammation of the esophagus. Behavioral symptoms are difficult swallowing and regurgitation accompanied by muscle spasms.

ESTIVATION: Prolonged dormancy in periods of high ambient temperature, characterized by reduced body temperature and metabolic activity. *Compare:* Hibernation.

ESTROGENS: A group of hormones produced in the ovarian follicles, but also found in the placenta, adrenal cortex, and testes, which promote the development of female secondary sexual characteristics, regulate estrus behavior, and sensitize the uterus to progesterone. Estrogens also sensitize the uterus of pregnant females to oxytocin.

ESTROUS BEHAVIOR: *See* Estrus.

ESTROUS CYCLE: A recurrent sequence of phases involving changes in sexual receptivity occurring in mature, nonpregnant females of many mammalian species, controlled by cyclic changes in hormone levels (e.g., follicle stimulating hormone, estrogen, luteinizing hormone, progesterone). One complete estrus cycle consists of four phases: proestrus, estrus (or true estrus), metestrus, and the period between two consecutive estruses, called diestrus. Typical duration of one complete estrus cycle is 21 days in mares,

20 days in cows, 16 days in ewes, and 21 days in sows. *Also see* specific estrous cycle—ELONGATED.

ESTROUS CYCLE, ELONGATED: *See* Elongated Estrous Cycle.

ESTROUS PERIOD: The part of the estrous cycle during which sexual behavior is manifested.

ESTROUS SYNCHRONIZATION: Coincidence of estrus (natural or artificially induced) occurring among a larger number of females than one would expect if the females were cycling randomly with regard to each other.

ESTRUS: A temporary state of sexual receptivity in female mammals occurring in coordination with ovulation. Behavioral signs of estrus, which may vary between species, typically include increased motor activity, higher excitability, reduction in feeding time, more frequent urination, presenting toward males or sexually cooperative females, tolerance of bodily contacts, and mounting. The most reliable sign of estrus is standing when mounted. The usual duration of estrus is 2-7 days in mares, 8-18 hours in cows, 1-2 days in ewes, and 1-3 days in sows. *Also see* specific estrus—FALSE; PSYCHOLOGICAL; QUIET; SILENT; SPLIT; STANDING; TRUE.

ESTRUS DETECTOR: Any person, animal, or device engaged in or used for identification of estrous females.

ET-EPIMELETIC BEHAVIOR: A variety of behavioral activities manifested to solicit attention, care, or help from other individuals. Synonym: Care-soliciting Behavior.

ETHICAL THEORY: Theoretical concepts of moral philosophy concerned with the principles of how to distinguish between ethically right or wrong human actions. Concern for animal well-being extends the applicability of ethical theories to judgements about human treatment of animals. The most commonly applied ethical theories in the assessment of treatment of farm animals are divine command, animal rights, utilitarianism, and rational egoism.

ETHICS: A subdiscipline of philosophy which deals with the question of human moral obligations. The study of moral standards and moral conduct.

ETHOGENETICS: *See* Behavioral Genetics.

ETHOGRAM: A record of behavioral activities.

ETHOLOGY: The study of the behavior of animals. *Also see* specific ethology—APPLIED; COMPARATIVE.

ETHOMETRY: The scientific discipline concerned with measurement and statistical analysis of behavior.

ETHOPATHY: Behavior indicative of or causing a pathological condition.

EUDAMONIA: Lasting well-being.

EUSTRES: A stress that is beneficial or that possesses some positive quality for the organism.

EUTHANASIA: The practice of causing easy death and terminating the irremedial suffering of an organism. Euthanasia is classified as active euthanasia if the subject is killed, or passive euthanasia if the death of the subject is not prevented.

EVENT, BEHAVIORAL: *See* Behavioral Event.

EVIDENCE: *See* specific evidence—ANECDOTAL; EXPERIMENTAL; SCIENTIFIC.

EVOLUTIONISM: A theory maintaining that existing species originated from primitive common ancestors and diversified as a consequence of natural selection. Antonym: Creationism.

EXAFFERENT STIMULATION: Mechanical stimulation caused by factors outside the organism's body.

EXALTATION: An extreme and pathological increase in the functioning of an organ(s).

EXCAVATION: Making a depression or a hole in the ground for dust-bathing, resting, cooling, nesting, or other purposes.

EXCITABILITY: Sensitivity of a tissue, organ, or organism to stimulation.

EXCITATION: Activation of neural centers or response organs.

EXCITATORY DRIVE MECHANISM: A hypothetical psychophysiological mechanism that controls drive.

EXCITATORY FIELD: An area of the brain activated by specific sensory input.

EXCITEMENT: A behavioral state characterized by an elevated level of arousal. Signs of excitement vary among species and different age groups and may include sensory focusing (orienting response), alteration of body posture, immobility, increased respiratory rate and heart beat, approach toward or retreat from the cause of excitement, emission or suppression of vocal signals and, in some circumstances, threat, attack or display of frustration.

EXCITEMENT, MIGRATION: *See* Migration Excitement.

EXCRETA: Residue of digestive and metabolic processes excreted from the body.

EXEMPLAR: An animal, trained to perform correct operant responses, provided in a test situation as a behavioral model to enable observational learning by naive individuals.

EXERCISE: Activity, often repetitive, that provides a psychophysiological challenge to an animal and promotes its maintenance and development. The activity may be rewarding to the animal in itself.

EXERCISE AREA: A designated area of housing facilities where animals can voluntarily exercise or are forced to exercise.

EXERTIONAL MYOPATHY: A spectrum of clinical conditions, interchangeably called tying-up, Monday morning disease, azoturia, and paralytic myoglobinuria, related to defects in skeletal muscle cells, or events causing damage to them (e.g., exercise or hard work after a few days of rest). Behavioral signs may include shortened stride; reluctance or inability to turn, jump, or move swiftly; stiffness; rigid posture; and hyperventilation.

This condition may be accompanied by excessive sweating and reddish or dark-brown discoloration of the urine by myoglobin.

EXHALATION: Expiration of air together with airborne substances. Antonym: Inhalation.

EXHAUSTION: A state of extreme fatigue.

EXISTENTIALISM: A theory espoused by Jean Paul Sartre which concerns human existence. Its central tenet is that the existence of human beings does not conform to any pre-existing pattern. According to this view, humans make themselves what they are.

EXOGENOUS RHYTHM: A biorhythm activated by events originating outside the organism. An exogenous rhythm requires periodic external input to maintain its oscillation.

EXOGENOUS STIMULATION: Stimulation arising from events occurring outside an organism's body.

EXOTHERM: An organism that derives its body heat primarily from external environmental sources such as sunlight.

EXPERIENCE: A summary of information retained at any given time from previous events in an organism's life. The term is also used to refer to ongoing stimulation as perceived at the time by a subject.

EXPERIMENT, YOKED: *See* Yoked Experiment.

EXPERIMENTAL BEHAVIORAL OBSERVATIONS: Behavioral observations which are collected in experimentally manipulated conditions to detect the effect(s) of a selected variable(s) on behavior, while controlling other variables. Antonym: Field Behavioral Observations.

EXPERIMENTAL CONDITIONING: Conditioning which occurs as a result of an experimental situation designed to form, strengthen, weaken, or extinguish a particular stimulus-response, or response-stimulus association. The two major forms of experimental conditioning are called, most commonly, classical and operant.

EXPERIMENTAL EVIDENCE: Evidence based on observations obtained by means of a defined experimental procedure. It is expected to be reproducible whenever the identical experimental procedure is repeated.

EXPERIMENTAL NEUROSIS: Behavioral disorders evoked by experimentally imposed difficult discriminatory problems or impossible tasks when failure to respond or an incorrect response results in severe punishment. Behavioral symptoms consist of stereotyped, inhibited, or compulsive behavior; chaotic movements or locomotion; frequent defecation; intense grooming, etc.

EXPIRATION: A segment of the respiration cycle characterized by contraction of the chest and lungs causing expulsion of air.

EXPLANATION, MECHANISTIC: *See* Mechanistic Explanation.

EXPLORATORY BEE: Any bee which searches for new resources of food and communicates discoveries to its hivemates.

EXPLORATORY BEHAVIOR: Behavior characteristically manifested during exposure to novel environments. The activities involved are usually increased overall alertness, sensory focusing, and locomotion accompanied by investigation.

EXPRESSION: An observable sign(s) indicative of some ongoing internal process.

EXPRESSIVE MOVEMENT: *See* Expression.

EXPULSION STAGE: *See* Labor.

EXTENDING, NECK: *See* Neck Extending.

EXTENSION: In a behavioral sense, straightening movement of an organism (or limb of an organism) which increases the angle between two segments of a joint. Antonym: Flexion.

EXTENSION (horse): Exaggerated forward movement of the legs in horse gaits.

EXTENSIVE HUSBANDRY: An animal production system characterized by low spatial density of animals, in which animals spend considerable amounts of time outdoors and obtain much of their feed by grazing or foraging. Such a system is often labor intensive and has a low level of mechanization.

EXTENSOR: Any muscle which increases the angle between two joint elements of a limb. Antonym: Flexor.

EXTENSOR REFLEX: Any reflexive extension of a limb. Such a reflex may occur in conjunction with a flexor reflex in the opposite limb.

EXTERIOR: Situated on the outside, or toward the outside.

EXTERNAL EAR: *See* Ear.

EXTERNAL ENVIRONMENT: The total exogenous surroundings, including social milieu.

EXTERNAL INHIBITION: Inhibition of a conditioned response when a novel stimulus occurs simultaneously with the conditioned stimulus.

EXTEROCEPTIVE NERVOUS SYSTEM: Part of the somatic nervous system which transmits afferent impulses from stimuli occurring outside the body. *Compare:* Proprioceptive Nervous System.

EXTEROCEPTOR: A sensory receptor activated by stimuli originating outside the body.

EXTINCTION: Suppression of a conditioned response by dissociation of reinforcement with the response, replacement of positive reinforcement with negative reinforcement (or vice versa) or establishment of an alternate conditioning regime in the same stimulus context.

EXTIRPATION: Removal or deactivation of an organ in order to study its effect on behavior.

EXTRACONSCIOUS: *See* Subconscious.

EXTRAPYRAMIDAL SYSTEM: That portion of the motor nervous system that does not include the corticospinal tract or reflex motor pathways.

EXTRINSIC BEHAVIOR: Behavior which does not have a specific response mechanism and can be performed in various ways.

EXTROSPECTIVE OBSERVATIONS: Observations of phenomena occurring outside the observer. Antonym: Introspective Observations.

EYE: A sensory organ specialized to receive and process light waves and transmit visual signals to the brain. The principal receptors are rods for scotopic vision and cones for photopic vision.

FACIAL EXPRESSIONS: Species-specific actions of muscles controlled by facial nerves, indicative of emotional state and possessing communicative properties. Facial expressions are most refined in primates.

FACIAL NERVE: The seventh cranial nerve which innervates facial muscles and mediates taste sensation from the cranial part of the tongue.

FACILITATION: A process or action by which something is made easier or more convenient.

FACILITATION, SOCIAL: *See* Social Facilitation.

FACTITIOUS: Unnatural or developed through artificial means.

FACTOR, CAUSAL: *See* Causal Factor.

FAINTING: A temporary loss of consciousness. Fainting can occasionally occur in young stallions during initial copulatory experiences.

FALSE ESTRUS: Behavioral signs of estrus during pregnancy.

FAMILIARIZATION: Acquisition of information by an organism about stimuli in its surroundings that facilitates recognition of these stimuli upon their reoccurrence. *Also see* specific familiarization—SOCIAL.

FAMILY: A collection of closely related items. In a social context, family refers to a group of related individuals. A family can be further classified as a nuclear family (e.g., parents and offspring) or extended family (e.g., associations of related sets of parents and offspring). The term also refers to a specific category of the taxonomic scale which subsumes groups of related genera.

FARM ANIMAL: A domesticated animal kept for economic purposes.

FARROWING (swine): The process of parturition.

FARROWING CRATE: *See* Crate.

FARROWING HYSTERIA: Parturient psychosis in swine.

FAR SIDE (horse): The right side of a horse.

FATHER: *See* Sire.

FATIGUE: The condition of being very tired and having reduced ability as a result of exertion or lack of rest. *Also see* specific fatigue—CAGE LAYER; MUSCLE.

FAULT: A failure in procedure or apparatus adversely influencing reliability of behvioral results.

FAUNA: All animal life present in a given region or area.

FEAR: An emotion caused by perception of danger. Behavioral symptoms of fear may include temporary immobility, hiding, escape, attack, urination, defecation, increased heart rate, etc. *Also see* specific fear—CONDITIONED.

FEATHER PECKING (poultry): Any pecking of the plumage, regardless of whether it is conducted by the bird itself or by another bird(s). Excessive feather pecking can cause structural damage to the plumage, and may be an indication of frustration or inadequacy of flock management.

FEBRICITY: The state of being feverish.

FECUNDITY: Capacity to reproduce, measured by number of surviving offspring. *Compare:* Fertility.

FEED, COMPLETE: *See* Complete Feed.

FEEDBACK: Any regulatory connection where the output affects the input. In a behavioral context, this term refers to processes in which the performance of an activity affects the continuation of the activity. *Also see* specific feedback—NEGATIVE; POSITIVE.

FEEDFORWARD: Any regulatory connection in which the rate of an ongoing process or activity is controlled in anticipation of the effects of the process or activity, before such effects actually have feedback influence. In animals, feedforward control appears to operate when feedback regulation would be too slow to prevent disruption in physiological homeostasis.

FEEDING: Delivery of feed to animals. *Also see* specific feeding—CONTROLLED; FORCED; INTRAVENOUS; LIMITED; RESTRICTED.

FEELING: Any subjective experience resulting from perceived sensations. The term is used to refer to mental states and emotions. It is also a subjective interpretation of exogenous and endogenous stimuli, primarily in relation to tactile sensation and pain.

FEIGNING BEHAVIOR: Behavior performed in an attempt to deceive (e.g., feigning of injury by a bird to distract a predator from its nest).

FELATIO: Oral stimulation of the penis.

FELINE: Cats or pertaining to cats.

FEMALES: *See* specific females—MONOESTROUS; POLYESTROUS.

FEMALE TENDING: *See* Guarding.

FEMINIZATION: Acquisition of female behavioral or body conformational characteristics by a male.

FERAL ANIMAL: Any animal that lives in a wild state, but whose progenitors have undergone a domestication process.

FERTILE: In a reproductive context, a state in which gametes are produced.

FERTILITY: Reproductive capacity. *Compare:* Fecundity.

FERTILIZATION: The process by which a sperm and ova join to form a zygote.

FETAL MOVEMENT: Detectable movement of the fetus or fetal organs during intrauterine development.

FETAL PRESENTATION: The particular position of the fetus when entering the birth canal during delivery.

FETAL RESORPTION: Absorption of a fetus by the maternal organism.

FETICIDE: Killing of fetuses; induced abortion. Male harrassment of pregnant females that have been inseminated by other males, apparently to cause abortion (natural feticide) and thus reduce the genetic fitness of rival sires, has been observed. Natural feticide has been suggested to occur fairly commonly in wild or feral horses.

FETLOCK: The area around the joint between the metatarsal bone and the first phalanx.

FETUS: A prenatal organism in the phase extending from the time that all major body structures are formed until parturition or hatching. *Compare:* Embryo.

FEVER: Elevated body temperature as a result of an increase in the set point level of the body's thermostatic control mechanism. Fever, generally, is a defensive response to pathogenic infection. *Also see* specific fever—MILK.

FIBER: Tissues or components of tissues that have a very elongated structure, e.g., nerve axon and dendrites.

FIELD BEHAVIORAL OBSERVATIONS: Observations of the behavior of animals in their natural habitat. Antonym: Experimental Behavioral Observations.

FIGHT: An aggressive social interaction involving interchange of forceful or potentially harmful actions, generally through some means of physical contact. Habitual fighting is considered to be a dangerous vice.

FIGHTING STANCE (chicken): A postural display indicative of readiness to attack and characterized by lowered head, extended neck, raised hackles, slightly spread wings in trailing position, and body positioned for a leap.

FIGURE EIGHT (horse): A dressage maneuver in which the horse follows two joint voltes or two circles, changing the direction of the circular path at the point where the voltes or circles join.

FILIAL: Pertaining to offspring, or to the generation of offspring as a whole.

FILIAL BEHAVIOR: Behavior characteristic of the relationship of offspring to parents, particularly behavior that induces parental assistance and cooperation, or requires parental leadership or parental training of offspring.

FILICIDE: Killing of one's own offspring. Filicide of neonates is thought to be caused by lack of parental experience in conjunction with some serious environmental deficiency or disturbance. *Compare:* Infanticide, Cannibalism.

FILLY (horse): A young female horse.

FILTERING, STIMULUS: *See* Stimulus Filtering.

FILTERING MECHANISM: A hypothetical mechanism of the perceptual system that, at any given time, determines selective attention to particular stimuli.

FINISHER DIET: A diet specifically formulated to prepare animals for marketing.

FISSION, COLONY: *See* Colony Fission.

FITNESS: *See* Genetic Fitness.

FITTING: Preparation of an animal for some event such as showing or sale. This process involves special feeding and grooming, coupled with training and exercise.

FIVE FREEDOMS: The Brambell Report proposed that all farm animals should have the freedom to perform five basic actions. These are: (1) to get up, (2) to lie down, (3) to turn around, (4) to stretch limbs, and (5) to groom.

FIVE-GAITED: A horse trained to execute five gaits (walk, trot, canter, slow gait, and rack) mainly for showing. In the American Saddlebred horse, five-gaited horses can be distinguished from three-gaited horses by their flowing manes. *Compare:* Three-gaited.

FIXATION: A process whereby a behavioral action or pattern becomes stereotyped and resistant to change.

FIXED ACTION PATTERN: Any action pattern typical of a given species or breed that is performed in a very similar way by its individual members. In contemporary ethology, the term fixed action pattern often is replaced by modal action pattern because of inevitable individual variations in behavior.

FIXED INTERVAL REINFORCEMENT SCHEDULE: A reinforcement schedule in which the reinforcing stimulus is never presented before a certain fixed length of time after the previous reinforcing stimulus. For example, if the interval is established as 10 minutes only the first response performed after the end of a 10-minute interval following the most recent presentation of the reinforcing stimulus will be reinforced. *Compare:* Variable Interval Reinforcement Schedule.

FIXED RATIO REINFORCEMENT SCHEDULE: A reinforcement schedule in which the presentation of the reinforcing stimulus is contingent upon completion of a fixed number of responses. If the ratio is 1:10, for instance, the reinforcing stimulus only occurs after each tenth response. *Compare:* Variable Ratio Reinforcement Schedule.

FLAPPING, EAR: *See* Ear Flapping.

FLAT RACING (horse): Racing on level terrain. *Compare:* Steeplechase.

FLAVOR: A quality of any substance perceived through the sense of taste.

FLEECE: The wool of a sheep.

FLEECE PULLING (sheep): An abnormal form of ingestive behavior mani-fested as pulling, tearing, and swallowing of the fleece of peers.

FLEHMEN: *See* Lip-curling.

FLEXIBILITY: Capability of an organism to move through a range of bodily positions without injury. In a broader sense, it may refer to an organism's ability to adjust to a changing environment.

FLEXION: Bending movement of an organism (or limb of an organism) that re-duces the angle between the segments of a joint. Antonym: Extension.

FLEXOR: A muscle that flexes two joint elements of a limb. Antonym: Extensor.

FLEXOR REFLEX: Any reflexive retraction of a limb.

FLICKING, EAR: *See* Ear Flicking.

FLIGHT: An escape response. Also, the manifestation of flying. *Compare:* Flying. *Also see* specific flight—NUPTIAL.

FLIGHT DISTANCE: The critical distance at which an organism will make an escape response upon the approach of another organism or object.

FLIGHT REACTION: Swift escape when an object perceived to be dangerous emerges within the limit of critical distance to the reacting organism.

FLIP-OVER (chicken): Sudden death syndrome in broilers apparently caused by heart failure. It occurs usually after the third week of life and most often affects fast growing individuals. Dead birds typically are found lying on their backs. (colloquial term)

FLOCK: A socially coordinated group of birds or sheep.

FLOCKING: The formation and maintenance of socially coordinated groups in birds and sheep. *Compare:* Aggregation.

FLOODING: An induced desensitization as a consequence of extended stimu-lation by an aversive stimulus.

FLOOR PEN: A pen where animals can ambulate over most or all of the floor area. The floors of such pens may be slatted concrete, solid concrete, or solid concrete covered with bedding material, e.g., wood shavings, cut straw, etc.

FLUSHING: Provision of an increased ration of food energy to a female for a short period before breeding. Flushing may help increase conception rate and number of offspring born. (colloquial term)

FLYING: Locomotion of an airborn organism in which lift and thrust derive from the action of wings.

FOAL HEAT (horse): The first postpartum estrus which, in most mares, occurs between 7 and 10 days after foaling.

FOALING (horse): Natural parturition in horses.

FOCAL ANIMAL: An animal that is chosen as the subject of behavioral observation while it remains in its group. The behavior of group members other than the focal animal is not recorded.

FOLLICLE STIMULATING HORMONE (FSH): A hormone produced by the anterior lobe of the pituitary gland that stimulates follicular growth and spermatogenesis.

FOLLOWERSHIP: A tendency by group members to accept control of their actions and direction of their behavior by others.

FOLLOWING RESPONSE: A response of an organism to the movement of another organism or object whereby the former travels at some relatively proximate distance behind the latter.

FOOD ANIMAL: A farm animal kept for production of food.

FOOD CALLING (chicken): *See* Tidbitting.

FOOD-RUNNING (poultry): Distinctive rapid locomotion most frequently displayed by young birds in the presence of their peers after grasping worms or worm-shaped objects with their beaks. A bird running with such an object in its beak is followed by peers who attempt to steal the object.

FOOD SELECTION: Preferential (and often species-specific) consumption of certain foods. Animals can be subdivided into herbivores, carnivores, and omnivores, according to the type of food consumed, and into specialists and generalists, according to the range of food consumed.

FOOT-AND-MOUTH DISEASE: A highly infectious virus disease of cattle, sheep, goats, and pigs. Behavioral symptoms are loss of appetite accompanied by rise in body temperature, reduction and progressive inhibition of locomotion, increased salivation, protrusion of the tongue, and occurrence of blisters on the feet, mouth, tongue, and teats.

FOOTROT: A disease occurring most often in sheep, but also in cattle and pigs. Behavioral symptoms are progressive lameness and swelling above the coronet followed by an odorous discharge from the affected areas.

FORAGE: A roughage used for feeding animals.

FORAGER BEE: Any bee that specializes in delivery of food to the hive. *Compare:* Exploratory Bee.

FORAGING BEHAVIOR: Behavior involving search for and intake of food.

FORCED COPULATION: Copulation after an unsuccessful attempt by a female to avoid mounting and intromission. In feral horses a stallion performing a forced copulation is occasionally aided by female members of his harem who actively prevent the avoidance or escape of the target female.

FORCED FEEDING: Any artificially induced delivery of feed into an organism's digestive tract.

FOREBRAIN: The largest part of the vertebrate brain, consisting of cerebral hemispheres, olfactory bulbs, basal ganglia, thalamus and hypothalamus.

The forebrain develops from the anterior part (prosencephalon) of the three divisions of the embryonic brain.

FORGET: To fail to retain information in memory due to disease, interference, or trauma to the brain. Also, inability to recall information stored in memory.

FORGING (horse): Defective leg action when the hoof of the hindleg hits the foreleg on the same side of the body.

FORM: The shape or appearance of an object, without regard to its nature or internal quality.

FORMULA: A set of symbols used to express in a concise way some functional or structural relationship.

FORTUITOUS: Accidental or due to chance.

FOSTER MOTHER: *See* Allomother.

FOSTERING: Promotion of growth and development. In a behavioral sense, the term refers to epimeletic behavior toward alien young, enabling their survival.

FOUNDER: To go lame or become unable to walk due to some physical disorder, such as laminitis. (colloquial term).

FOVEA: An area of the retina that is densely packed with cones. Vision is most acute in the portion of the visual field that is focussed onto a fovea.

FOWL: Poultry and related wild species of birds.

FOWL CHOLERA: A contagious disease of poultry caused by *Pasteurella multocida*. Behavioral symptoms are severe diarrhea, dropped wings, loss of appetite, discolored combs and wattles, eye discharge, and rapid respiration accompanied by high body temperature.

FOWL PEST: *See* Newcastle Disease.

FOX HUNTING (horse): A popular cross-country horseback recreational activity.

FOX-TROT (horse): A medium speed (8-11 km/hr) four-beat gait comfortable for the rider. The diagonal legs are raised at the same time, but the forehoof contacts the ground before the hindhoof.

FRACTIONATION, DIET: *See* Diet Fractionation.

FRASER DARLING EFFECT: A phenomenon in which production of offspring is highly synchronized within a population so that the capacity of predators to take the offspring as prey is surpassed. Under the circumstances in which this phenomenon occurs, it is postulated that such a reproductive strategy is adaptive because, in the long term, fewer offspring would be lost to predation.

FRATERNAL TWINS: Twins originating from two fertilized ova.

FRATRICIDE: Killing of siblings.

FREE-CHOICE: A type of feeding in which an animal is free to choose among two or more alternatives.

F

FREEDOMS, FIVE: *See* Five Freedoms.

FREE GOING (horse): A horse that performs a gait in an easy and seemingly effortless manner.

FREE-LEGGED PACER (horse): *See* Natural Pacer.

FREEMARTIN: The female twin of a bull calf. Freemartins are usually infertile.

FREE-RUNNING RHYTHM: A biorhythm displaying spontaneous cycling pattern.

FREE STALL: A stall within a housing unit which animals can enter and exit freely. Free stalls are commonly part of loose housing systems for dairy cows.

FREEZING BEHAVIOR: Adoption of a fixed, immobile stance. Freezing may be manifested by animals that perceive danger, and may be related to fear. It also may be shown by predators hunting prey. In either case, the behavior is thought to render an animal less susceptible to detection.

FRESH: Referring to a postpartum cow. (colloquial term)

FRESHEN (cattle): Parturition and commencement of lactation. (colloquial term)

FROG POSTURE: Full or partial extension of both hind legs forward along the sides of the recumbent body. In cattle and horses, frog posture is an indicator of hip dislocation, rupture of the adductor muscles, or paralysis of the obturator nerve.

FROLICKING: A form of play behavior, observed particularly in young animals and often in a social context.

FRONTAL: Referring to or located in the forehead.

FRONTAL LOBE: The anterior part of the cerebral hemisphere.

FRUSTRATION: A theoretical concept referring to a condition that is produced when an organism is blocked in its attempts to achieve a goal. Behavioral symptoms of frustration vary among species and may include elevated levels of pacing, grooming, preening, pawing, vocalization, and aggression. Sustained frustration promotes development of stereotypy or apathy.

FULL-PASS (horse): An artificial maneuver in which the horse moves sideways in a direction 90 degrees from its median lane. *Compare:* Half-pass.

FUNCTION: In a behavioral context, that which is accomplished by a given action. In a statistical sense, a function mathematically defines the relationship between two or more variables.

FUNCTION, DEMAND: *See* Demand Function.

FUNCTIONAL ANALYSIS: An analysis that focusses on the determination of the consequences of a given behavior.

FUNCTIONAL DISORDER: A disorder affecting the function but not the structure of an organ.

FUNCTIONALISM: A school of thought that considers all psychological processes within the context of natural selection and, therefore, examines them according to their function and adaptive value.

FUNCTIONAL UNIT: A molar category subsuming behavioral actions that serve a common purpose, but may be performed in a different way. For example, a functional unit identified as "eating" may include intake of roughage, chopped roughage, mixed diet, concentrates etc., each such form of feed being consumed in a slightly different way.

FUSION: Inseparable combination of two or more stimuli.

GABBLING (geese): One of the vocalizations of geese, consisting of repeated short, cackling sounds and most typically produced when encountering familiar peers.

GAIN, AVERAGE DAILY: *See* Average Daily Gain.

GAIT: Action of the legs during locomotion. *Also see* specific gait—DIAGONAL; LATERAL.

GALLOP: The fastest naturally developed gait. The leg movement is similar to the canter, except the diagonal beat is extended so the forehoof hits the ground slightly later than the hindhoof and the phase of total suspension in the air is longer (a fast gallop is a distinctly four-beat gait).

GALVANIC SKIN RESPONSE: A change in the electrical resistance of the skin resulting from emotional arousal, tension, fear, or pain.

GAMBOLING: Bouncing and turning, seemingly stiff-legged, in the air. Gamboling can be observed mainly in the young playing sheep and to a lesser degree in goats.

GAMETE: A reproductive cell having a haploid set of chromosomes.

GANDER: A mature male goose.

GANGLIA, BASAL: *See* Basal Ganglia.

GANGLION: A conglomerate of nerve cells or cell bodies outside the central nervous system.

GAS: *See* General Adaptation Syndrome.

GASPING: Difficult breathing characterized by abrupt and often great effort in drawing breath.

GASPING TURKEY SYNDROME: Gasping accompanied by spreading of wings and often unsteady gait, occurring most commonly in heavy turkeys after 10 weeks of age. Turkeys predisposed to gasping appear to have lowered tolerance to excitement and exercise and are prone to sudden death when moved or during handling and semen collection.

GATE, TEXAS: *See* Texas Gate.

GATHERING AREA: A location where animals frequently aggregate, usually near water sources, feeders, shelters, or gates. Gathering sites are easily recognizable by damaged vegetation, trampled ground, and accumulations of excreta. The term also is used to refer to locations within animal holding facilities designed to collect groups of animals.

GELDING: A castrated male horse.

GENE: A basic unit of hereditary transmission possessing coded information for synthesis of specific peptides.

GENEALOGY: The study of the genetic origin of an organism or group.

GENE LOCUS: The site on a chromosome occupied by a given gene.

GENERAL ADAPTATION SYNDROME (GAS): A concept proposed by Hans Selye referring to the sequence of physiological changes in an organism exposed to a stressful situation. There are three consecutive phases of GAS: alarm reaction, stage of resistance, and stage of exhaustion.

GENERALIZATION: Responsiveness to stimuli that approximate the original stimulus but are not identical to it.

GENERALIZATION GRADIENT: A decrease in strength of response as a function of decreasing similarity between the test stimulus and the original stimulus.

GENERALIZED STIMULUS: A stimulus to which an organism responds at an intensity that is based on the perceived similarity or identity of this stimulus with an original stimulus.

GENERATION: Contemporary individuals of a species that are of similar age; individuals that belong to the same level in a genealogical sequence.

GENERATOR POTENTIAL: A graded electrical potential produced in a receptor cell from the transduction of environmental energy associated with some stimulus. If the generator potential is sufficiently great, the receptor produces an action potential that is transmitted along afferent neural pathways. The information inherent in the sense datum is represented by the generator potential.

GENETIC DISPOSITION: *See* Hereditary Predisposition.

GENETIC DRIFT: Change in population gene frequency due to chance (sometimes also called random genetic drift).

GENETIC ENGINEERING: The application of techniques based on principles of molecular biology to modify reproduction, metabolism, growth, or behavior of animals.

GENETIC FITNESS: The relative contribution of a particular genotype to the gene pool of the subsequent generation. Individual fitness is measured by the number of an organism's offspring that reach sexual maturity. Inclusive fitness is the sum of the individual's own contribution through its offspring and the contribution of kin by genes identical by descent.

GENETICS: The scientific discipline concerned with hereditary transmission of biological characteristics. *Also see* specific genetics—BEHAVIORAL.

GENETIC STOCK: A general term for any population of individuals considered to be genetically distinct from other such populations.

GENITAL: Of or pertaining to the organs of reproduction.

GENOTYPE: The genetic constitution of an organism.

GENUS: A taxonomic category that incorporates a group of related species. Related genera are, in turn, incorporated into a family.

GEOPHAGIA: Eating soil or earthy substances.

GEOTAXIS: Taxis related to gravity.

GEOTROPISM: An orienting response to gravity.

GERONTOLOGY: The study of aging processes.

GESTALT PSYCHOLOGY: A psychological school that stresses that behavior cannot be simply analyzed in an atomistic way because components of behavior derive their nature and role from the whole. The components cannot be fully understood apart from the whole, nor can the whole be understood as a simple summation of its parts. Psychological processes are considered to be the result of complex interactions between the totality of elements making up or influencing the organism.

GESTATION: The process of intrauterine development of a mammalian organism.

GESTATION CRATE: *See* Crate.

GESTATION, PROLONGED: *See* Prolonged Gestation.

GESTURE: A movement or expression displayed for the purpose of visual communication.

GETAWAY CAGE: A cage for laying hens, designed to provide higher environmental complexity than battery cages. In addition to water and feeder, these cages also have a nesting box, sand-bathing box, and roosts.

GILT: A young female swine up to the time of first farrowing.

GLAND: An organ specialized to secrete substances to be used by other organs or to be excreted from the body.

GLANDULAR RESPONSE: The reaction of a gland to efferent stimulation.

GLOSSOPHARYNGEAL NERVE: The ninth cranial nerve, which innervates muscles of the pharynx and mediates immediate taste sensations from the caudal parts of the tongue.

GLUCOSTATIC THEORY: A theory that proposes that hunger and satiation reflect an organism's endeavour to maintain a constant blood glucose level.

GLYCOGEN: A polysaccharide that serves as a body reserve of energy necessary for activity of muscles and other tissues.

GNAWING, WOOD: *See* Wood Gnawing.

GOAL: A commodity or circumstance for which an animal has both mental representation and intention to acquire or achieve. An incentive.

GOBBLING (turkey): Vocalization of mature male turkeys, produced usually as a series of 3-7 sounds, accompanied in most cases by moderate head movements.

GOLGI-MAZZONI CORPUSCLES: Encapsulated nerve endings in the skin regarded as receptors of temperature or pressure.

GONAD: A sexual gland called testis in males and ovary in females that produces gametes and sexual hormones.

GONADOTROPHIN: A compound that has the ability to stimulate gonadal activity. The term is a common designation for luteinizing hormone (LH), also called interstitial cell stimulating hormone (ICSH), and follicle stimulating hormone (FSH) from the anterior pituitary gland, as well as chorionic gonadotrophins found in urine of pregnant humans and some other primates, and gonadotrophins found in the plasma of pregnant horses.

GOOSE-STEPPING (swine): Unusually high action of the legs accompanied by slower locomotion; it may be a symptom of deficiency of pantothenic acid.

GOSLING: A young goose prior to sexual maturation.

GRADED SIGNAL: A signal whose information is determined both by the fact that it is emitted and by various forms of signal alteration, such as intensity, frequency, or duration. *Compare:* Discrete Signal.

GRADIENT: Rate of change of a stimulus or response (the slope of the stimulus or response curve at any given point).

GRASPING REFLEX: Reflexive holding onto an object. Reflexive grasping of the body of the mother is very common in neonates. Grasping also occurs in adults during copulation.

GRAVID: Referring to a female carrying an embryo or fetus; pregnant.

GRAZING: The act of consuming standing vegetation.

GRAZING DENSITY: *See* Spatial Density.

GREASY PIG DISEASE: A bacterial disease caused by *Staphylococcus hyicus*. Greasy pig disease occurs primarily in young piglets from hypogalactic and excitable dams because of bacterial infection through lesions caused by elevated levels of ear sucking among piglets. Greasy pig disease causes wrinkled skin, covered with greasy, gray-brown clumps, and begins behind the ears, subsequently spreading around the eyes and over the abdomen. Mortality of affected pigs is high. Behaviorally, there appears to be no indication of pruritus.

GREETING SIGNAL: Any auditory, visual, tactile, or other sign, or combination of such signs, displayed by an organism upon becoming aware of the presence of another organism and indicating acknowledgement of its presence and friendly intention toward it.

GREGARIOUS: Referring to an individual that characteristically lives in association with conspecifics in a group, herd, or flock. The same term may be applied in a general sense to a species whose individuals are typically gregarious. Antonym: Solitary.

GROAN (horse): A vocalization of approximately 0.5 second duration, having a broad sound frequency range. It is assumed to be indicative of discomfort and frustration.

GROAN (sheep): A low amplitude sound of intermediate duration (0.5-1.2 sec) produced with a slightly open mouth. Groans are characteristically emitted by animals that are ill, those apparently suffering from pain, or occasionally by healthy animals adjusting posture when recumbent.

GROOMING: An act of integumentary care (e.g., licking, scratching) to remove parasites, smooth ruffled fur, remove dirt, etc. Grooming is subdivided into: self-grooming (an animal grooming itself) and allogrooming (an animal grooming another animal). Grooming may also have a social function, appearing as conciliatory behavior (a subordinate animal grooms a dominant animal), reciprocal altruism (animals groom each other), or ritualized behavior performed during courtship or conflict.

GROOMING DANCE (bees): Rapid dorso-ventral shaking of the body that attracts other bees to clean bristles inaccessible to the bee itself.

GROUND-RUTTING: *See* Horning.

GROUND-SCRATCHING (poultry): A sequence of rapid foot movements performed to manipulate the ground surface. It occurs frequently during food gathering and to a lesser degree in frustrating situations.

GROUP: A collection of animals. Generally the term is applied to situations in which the animals are of the same species and the composition of the group is relatively stable over time. *Also see* specific group—HETEROGENEOUS; HOMOGENEOUS.

GROUP ACCEPTANCE: Reception of a new member into an established social group.

GROUP COMPOSITION: The makeup of a group as determined by the characteristics of its members; e.g., age, gender, reproductive condition, social status.

GROUP DISTANCE ZONE: An area surrounding a group of organisms defined by the distance the group seeks to maintain between itself and other groups. The size of this zone depends on circumstances existing at a given time, such as population density, group size, season, availability of feed resources. This term generally is used in the context of conspecific social interactions. *Compare:* Individual Distance Zone.

GROUP EXPERIMENT: Any experiment in which the primary objective is to evaluate the behavior of a group as a whole.

GROUPING: The formation of a group of animals by natural means (e.g., herd formation as a result of social attraction) or by human action, (e.g., allocation of a number of animals to a given pen or grouping of dairy cows according to milking performance).

GROUP ODOR: An odor characteristic of a group that assists mutual identification of group members (documented in bees and in some mammalian species).

GROUP SIZE DEPENDENCE: A phenomenon in which the expression of a behavior is affected by the number of animals in a group (e.g., time necessary to establish a social hierarchy). *Compare:* Density Dependence.

GROUP STRUCTURE: *See* Group Composition.

GROWTH HORMONE: *See* Somatotropin.

GROWTH IMPLANTS: Subcutaneously implanted biochemical substances that improve feed conversion and growth rate.

GROWTH PROMOTANTS: Additive substances mixed with the feed to improve feed conversion and growth rate.

GRUNT (sheep): A low amplitude sound of short duration (0.3-0.75 sec) produced with a closed or slightly open mouth. It is emitted during reestablishment of contact between peers and is indicative of greeting.

GRUNT (swine): The most common vocalization of mature swine produced as a sound of low to medium amplitude. Grunting may consist of single grunts, but more commonly occurs as a series of repeated sounds produced with mouth closed or only slightly open. The pitch of grunts is usually between 1 and 4 kHz. According to the duration of individual sounds they can be subdivided into short, medium, or long grunts. Short grunts (0.1-0.2 sec) appear to be a sign of mild excitement and often are produced when a pig is frustrated or greeting another individual. Mid-length grunts (0.2-0.4 sec) are often produced during interactions with familiar peers and also during the milk ejection phase of a normal nursing cycle of a sow. Long grunts (0.4-1.2 sec) are produced in response to tactile stimulation such as occurs during courtship and the nursing cycle (particularly the nosing and slow suckling phases). *Also see* specific grunt—COURTSHIP; NURSING.

GUARDING (cattle): The maintenance of close proximity by a bull to a cow in proestrus and estrus.

GUSTATION: Perception of a substance through detection of molecules derived from direct contact with the substance (taste). In mammals and birds, gustation is mediated by specialized receptor cells in the taste buds on the tongue.

GUSTATORY: Pertaining to the sense of taste.

GUTTURAL: Pertaining to the throat.

GYMKHANA (horse): Popular horseback games consisting of a combination of barrel racing, polebending, keyhole race, and stake.

GYNADOMORPH: An organism that possesses both male and female characteristics.

HABIT: A persistent pattern of behavior.

HABITAT: The surroundings and conditions in which an organism lives.

HABIT BREAKING: A conditioning process focussing on extinction of habitual forms of behavior.

HABITUATION: A simple form of learning involving a permanent reduction or elimination of a response in the absence of any overt reward or punishment. Habituation generally is distinguished from extinction on the basis that the response that is reduced or eliminated is unlearned.

HAEMATURIA: Presence of blood in the urine.

HAIRBALL: A ball of hair in the digestive tract, developed most frequently in bovines as a consequence of excessive integumentary licking.

HAIR CELLS: Cells shaped as hair-like protrusions. Hair cells are involved in the reception of sounds.

HAIR CHEWING: Chewing or nibbling on the pelage of conspecifics. It occurs in animals of all ages in free stalls or tie stalls, particularly if the animals are fed high concentrate diets.

HALF-PASS (horse): A dressage maneuver in which the horse moves on a 45 degree angle from its median lane. Half-pass is principally a Travers executed in a semi-lateral direction.

HALO EFFECT: The influence of one characteristic of a subject upon the rating of independent characteristics of the same subject.

HALOTHANE TEST: A test to detect the propensity of swine to develop porcine stress syndrome. It involves inhalation of halothane gas and results in muscular rigidity of pigs that are susceptible to factors causing the syndrome.

HALT (horse): A dressage position in which the horse stands motionless with apparently equal distribution of weight on all four legs. The neck is raised with the front of the head close to vertical and the ears indicating attention.

HALTER: A device made of leather straps or rope that closely fits the shape of an animal's head and commonly is used to tie and handle large farm animals.

HALTER-PULLING (horse): An action in which a tethered horse pulls backward powerfully on its halter. The horse may have its hindhooves slip, causing it to fall and roll, or if the halter breaks, the horse may tumble backward and risk injury. Halter-pulling may become habitual in some horses and is considered a dangerous vice.

HALTER SLIPPING: Intentional escape from a halter restraint. Some animals become skilled at halter slipping. To prevent this behavior, once it becomes a habit, specially designed halters must be used.

HAND BREEDING: Human-controlled and time-limited pairing of sexual partners for the purpose of mating.

HANDICAP (horse): A type of race in which competing horses must carry an individually assigned amount of weight, theoretically equalizing their chances of winning.

HAPLODIPLOIDY: A situation in which sex determination is based on the chromosomal complement of an organism, haploid individuals being male and diploid individuals being female. Some hymenopterans, including the honey bee, manifest haplodiploidy.

HAPLOID: Containing only one set of chromosomes. Gametes are haploid cells. *Compare*: Diploid.

HAPTIC: Referring to the sense of touch.

HAPTOMETER: An instrument for measuring touch sensitivity.

HARASSMENT: Persistent, intentional disturbance of an individual by another individual or group.

HAREM: A group of sexually mature females dominated by and mating with one male.

HARM: To cause injury or damage. In the context of animal welfare, the term generally refers to the reduction of an organism's quality of life, or prevention of achievement of a higher quality of life, by denial of its biological needs and other relevant interests or by subjecting it to excessive stimulation.

HARNESS RACING (horse): Racing in harness, pulling a sulky and driver. The gait used is the trot or the pace.

HATCHING: The process whereby an organism emerges from the egg laid by its dam.

HATCHING SYNCHRONIZATION: A phenomenon in which eggs incubated together all hatch within a short time period.

HAUTE ECOLE (horse): Advanced dressage focussing on refined execution of required movements.

HEAD DIPPING: A period of rhythmical submergings of the head and part of the neck in water, manifested by some waterfowl during courtship.

HEAD JERKING: Voluntarily controlled short, quick, single movements of the head.

HEAD NYSTAGMUS: Oscillatory head movement that occurs after the body of an animal has been subjected to rotation.

HEAD PRESSING: Extended and repeated leaning forward and pushing of the head against a wall, fence post, or other fixed objects. Head pressing commonly is a symptom of brain damage (due to, e.g., encephalomyelitis, hepatic encephalopathy or elevated intracranial pressure).

HEAD RESTRAINED PRESENTATION: Fetal presentation in which the anterior of the body approaches the birth canal, but the neck is bent and the head directed to the rear.

HEAD SCRATCHING: A comfort movement oriented toward the head area. Generally it is a response to skin irritation or a symptom of mild excitement.

HEAD SHAKING (poultry): A rapid series of side-to-side flicking movements of the head.

HEAD-SHY (horse): Refers to a horse that resists close approach to its head or having a bridle put on its head.

HEAD TOSSING: A vigorous and often repetitive vertical movement of the head. Head tossing often can be observed in horses under saddle or in unfamiliar environments. It is thought to be a sign of mild excitement or improperly fitting equipment.

HEALTH: The state of an organism's existence that is characterized by unimpaired biological functioning, complete physical and psychological adjustment to its surroundings, and uncompromised well-being.

HEAR: To perceive stimuli through the auditory sense mode. Audition.

HEART RATE: The number of contractions of the heart muscle in a given time, commonly one minute.

HEAT: A sensation resulting from simultaneous stimulation of thermal receptors in the skin. Colloquially, the word heat refers to estrus.

HEAT, STANDING: *See* True Estrus

HEAT DETECTOR: *See* Estrus Detector.

HEAT PATCH: An estrus detection device, attached to the top of a cow's rump, that is able to register pressure (e.g., by color change) induced when the cow is mounted by another animal.

HEAT PERIOD: *See* Estrus.

HEAT STROKE: A condition that occurs during a period of hot weather and is characterized by lethargy or complete inhibition of locomotion.

HEDONISM: A theory that the generation of one's own pleasure is the guiding principle of action.

HEIFER: A female bovine younger than 3 years that has not delivered a calf.

HELIOTROPISM: Phototropism in which sunlight is the orienting factor.

HELPLESSNESS, LEARNED: *See* Learned Helplessness.

HEMIPLEGIA: Paralysis or paresis affecting one-half of an organism's body.

HEMISPHERES, CEREBRAL: *See* Cerebral Hemispheres.

HEMOPTYSIS: Coughing up and sometimes swallowing blood from the lungs or lower air passages. Hemoptysis usually is caused by hemorrhage in the lungs. Repeated swallowing without drinking or eating may be an indication of such bleeding.

HEMORRHAGE: Arterial, venous, or capillary bleeding.

HEN: A sexually mature female turkey or chicken.

HEPATIC: Pertaining to the liver.

HEPATITIS: Inflammation of the liver. Behavioral symptoms may be irregular appetite, listlessness, and pendulous abdomen.

HERBIVOROUS: Referring to animal species which subsist on vegetation.

HERD: A socially coordinated group of ungulates. In an agricultural sense, a herd is a group of horses, cattle, goats, or swine considered as one managerial unit.

HERDING: The formation and maintenance of socially coordinated groups by mammals. *Compare*: Aggregation.

HERDING POSTURE: A stance adopted by stallions manifested as a lowered head and neck with ears laid back. It often is performed when the stallion drives females from one place to another. Rams and bucks may perform a similar behavior pattern but in a less elaborated form and less frequently.

HEREDITARY PREDISPOSITION: The genetic potential to develop certain characteristics in a specific environment, should the organism experience this environment.

HEREDITY: The transmission of characteristics from parent to offspring through information coded on genes.

HERITABILITY: *See* Heritability Coefficient.

HERITABILITY COEFFICIENT: A population-specific ratio between estimates of genetic variance and phenotypic variance calculated separately for each individual trait or characteristic.

HERMAPHRODITISM: The presence of male and female sexual organs in the same organism.

HETEROCHRONY: A difference in time or speed between two processes.

HETEROGENEOUS: Composed of different elements or originating from different sources.

HETEROGENEOUS GROUP: Two or more individuals differing in some basic characteristic (e.g., sex, reproductive stage, age, or experience). Antonym: Homogeneous Group.

HETEROGENEOUS SUMMATION, LAW OF: *See* Law of Heterogeneous Summation.

HETEROREXIA: An compelling appetite for atypical food.

HETEROSEXUALITY: Sexual attraction to individuals of the opposite sex.

HETEROSIS: Increased vigor associated with increased heterozygosity.

HETEROSIS EFFECT: *See* Heterosis.

HIBERNATION: A dormant stage occurring during winter in some animal species, characterized by a drop in body temperature and minimal metabolic activity. *Compare*: Estivation.

HIDDEN TESTES: *See* Cryptorchid.

HIDE: To engage in cryptic behavior.

HIDROSIS: Extreme sweating.

HIERARCHY: *See* specific hierarchy—COMPLEX; LINEAR; LINEAR-TENDING; SOCIAL.

HIERARCHY LOOP: Any nonlinearity in a dominance-subordinance ranking scale. This may occur in groups of three or more animals (e.g., animal one dominates animal two, animal two dominates animal three, but animal three dominates animal one).

HIGH-GAITED (horse): Referring to leg action in which there is high raising of the hooves from the ground and great flexion of the hocks.

HIGH PERMEABILITY OF JOINTS: *See* Hypermobility of Joints.

HIGH-STEPPING (chicken): A postural display of males characterized by semicircular movement performed with high, slow steps around an opponent. The bird holds its head and tail high, thrusts out its chest and slightly trails its wings. High-stepping is a threat signal.

HIGH-STEPPING (sheep): Elevated raising of the legs when walking. High-stepping may indicate ovine encephalomyelitis.

HINDBRAIN: The caudal part of the vertebrate brain, which consists of the cerebellum, pons, and medulla oblongata. The hindbrain develops from the posterior part (rhombencephalon) of the three divisions of the embryonic brain.

HINNY: Offspring of a stallion and a jenny.

HISSING (geese): A sibilant sound, up to several seconds long, produced as a threat signal by geese. This term also refers to similar activities in other species such as cats and some reptiles.

HISTOGRAM: Graphical presentation of a frequency distribution using rectangular bars. The bar represents the interval of a class and the height of the bar reflects the frequency in each class.

HISTOTROPH: The uterine secretions that provide necessary nutrients for the development of an embryo prior to formation of the placenta.

HIVE: A man-made housing for a colony of bees. Hive also refers to the colony of bees itself.

HOARDING: The storage of food or other materials in a cache(s) located within an animal's home range.

HOBBLE: Any leather band or rope placed around an animal's leg(s) to restrain it.

HOBBLES: *See* specific hobbles—BREEDING; MILKING.

HOCK TWITCH (cattle): A flexible loop placed above the hock. When tightened with a short stick, the loop compresses the Achilles tendon and prevents kicking.

HOG CHOLERA: A highly contagious viral disease of swine of all ages. Behavioral symptoms are reduced appetite, reduced ambulation accompanied by increased body temperature, trembling, leg paralysis (particularly of the hind legs), sticky eye discharge, constipation followed by diarrhea, and finally death.

HOGGET: A young sheep from weaning to first shearing.

HOLISM: A theory that the whole is more than just a summation of its individual parts. In application to behavioral science, this theory stresses that the more complex actions of an organism cannot be fully explained through isolated understanding of individual component actions.

HOMEOSTASIS: Maintenance of a state of psychophysiological balance within an organism by means of internal control mechanisms. These mechanisms operate within specialized systems, such as the extracellular fluid transport systems, respiratory system, digestive system, endocrine system, nervous system, musculoskeletal system, and reproductive system.

HOMEOSTATIC DRIVE: A drive resulting from a condition of psychophysiological imbalance.

HOMEOTHERMY: Maintenance of a relatively stable body temperature in a fluctuating environmental temperature. *Compare*: Poikilothermy.

HOME RANGE: A locality where an individual(s) conducts all its principal activities.

HOMING: A phenomenon in which an organism shows a tendency to return to its original home when transferred to another locality.

HOMOGENEOUS: Composed of identical or similar elements or elements originating from the same source.

HOMOGENEOUS GROUP: Two or more individuals identical or similar in some basic characteristic (e.g., sex, reproductive stage, age, or experience). Antonym: Heterogeneous Group.

HOMOLOGOUS BEHAVIOR: Behavior in different species that is similar in form, but not necessarily in function. The behavior is similar as a result of the species' common phylogenetic origin. *Compare*: Analogous Behavior.

HOMOSEXUALITY: Sexual attraction for individuals of the same sex.

HONEY GUIDES: Species-specific distinctive flower markings visible only in the ultraviolet spectrum and apparently adapted to attract insects that mediate pollen transfer. Honey bees can learn to identify such markings and search preferentially for them if the flowers so marked provide a good yield of nectar and pollen.

HONEYMOON FLIGHT: *See* Nuptial Flight.

HONKING (geese): One of the vocalizations of geese, consisting of rhythmical loud sounds. Honking is typically produced when geese are flying or excited.

HOOF RUBBING (horse): Chafing of the crown of one hindleg by habitual resting of the hoof of the other hindleg on it. This may lead to leg injury, particularly if the horse wears shoes that have sharp edges.

HORMIC THEORY: A theory that emphasizes the central role of purposive striving and instincts in the behavior and psychology of organisms.

HORMONE: A chemical substance produced in specialized cells or tissues that has a regulatory effect(s) on the activity of other cells, tissues, glands, or organs. Hormones also have profound influences on behavior, e.g., testosterone affects aggressiveness and sexual behavior.

HORNING (cattle): Digging or churning the ground using the horns or forehead. This type of behavior is generally displayed by males and is assumed to be a sign of excitement and a warning signal.

HORSE LAUGH: Lip-curling (Flehmen) in horses. (colloquial term)

HOT WALKER: A person or mechanical device that leads horses in a circle at a slow walk to cool them after training.

HOUSING: *See* specific housing—CONFINEMENT; INDOOR; LOOSE; OUTDOOR.

HUDDLING: Maintaining close bodily contact with other members of a group. The term huddling is used for situations in which the bodily contact is initiated voluntarily by the individuals involved.

HUE: The characteristic of visual sensations that results from stimulation of the eye by light of different wavelengths.

HUMANE: An attitude of kindness toward and sympathy for sentient organisms coupled with intention to prevent or minimize their suffering.

HUMANISM: A theory that rejects religious beliefs and advocates that the only intent of human action should be to promote human well-being.

HUNGER: An uncomfortable sensation induced by fasting. Hunger generally is associated with increased appetite for food, provided the duration of fasting has not resulted in lethargy.

HUNGER, SPECIFIC: *See* Specific Hunger.

HUNTER (horse): A horse trained, exhibited, or competing in the maintenance of an even hunter pace and good jumping style. Hunters are judged working natural obstacles such as hedges, stone walls, and board fences that are 1.0-1.5 m high.

HUNTING: Behavior conducted by carnivorous species to prey on other animals.

HUSK: A parasitic disease of cattle, sheep, and goats that affects the lungs. Behavioral symptoms are husky coughs and, in an advanced stage, difficult breathing and rigid posture with extended neck.

HYBRIDIZATION: Sexual reproduction in which gametes originate from animals of different species.

HYDROTAXIS: Taxis in response to water or humidity.

H

HYGROMA: Excessive swelling around joints.

HYPERACUSIA: Abnormally high auditory sensitivity.

HYPERAEMIA: Increased blood circulation visibly manifested by reddishness of the affected tissue.

HYPERAFFECTIVITY: Abnormally high general sensitivity. *Compare*: Hyperesthesia.

HYPERAGIA: *See* Hyperalgesia.

HYPERALGESIA: An extreme sensitiveness to pain.

HYPEREMOTIVITY: Abnormally high emotional reactivity.

HYPERESTHESIA: Abnormally high sensitivity to stimulation. This term is generally used in reference to specific sensory modalities (e.g., auditory, gustatory, olfactory, tactile, or visual hyperesthesia). *Compare*: Hyperaffectivity.

HYPEREXTENSION: Movement along the sagittal plane of an organism (or limb of an organism) that increases the angle between the segments of a joint beyond 180 degrees.

HYPERGALACTIA: Milk production by a postpartum female exceeding that required for raising offspring.

HYPERKINESIA: Abnormally high level of motor activity.

HYPERKINESIS: *See* Hyperkinesia.

HYPERMETRIA: Motor action exaggerated beyond what is necessary to reach an object or goal.

HYPERMOBILITY OF JOINTS: Unusually high elasticity of joints. In extreme cases, the flexibility does not permit upright posture or rising from recumbency. Hypermobility of joints occurs in some breeds of cattle (Jersey, Holstein-Friesian) and appears to be a congenital defect.

HYPEROREXIA: Pathologically strong appetite.

HYPEROSMIA: Abnormally high sensitivity to odors.

HYPEROXIA: Excessive supply of oxygen to body organs.

HYPERPHAGIA: Consumption of an excessive quantity of food. *Compare*: Polyphagia.

HYPERPNEA: Abnormally intense respiration characterized by deep pulmonary movements.

HYPERREFLEXIA: Exaggerated reflexes or excessive reaction to reflex-inducing stimuli.

HYPERSENSITIVITY: Abnormally high sensitivity to stimulation.

HYPERSEXUALITY: Abnormally high sexual excitability or generally high level of sexual activity.

HYPERSOMNIA: Pathological drowsiness.

HYPERSTHENIA: Abnormally high strength.

HYPERSTIMULATION: Excessive intensity or complexity of stimulation causing neurosis or harm to the subjected organism.

HYPERTHERMIA: Abnormally high body temperature.

HYPERTONIA: Abnormally high tonicity in skeletal muscles.

HYPERVENTILATION: Excessive flow of air into and out of the lungs resulting in reduced carbon dioxide levels in the blood.

HYPNO-: A prefix signifying a relation to sleep.

HYPNOLOGY: The science of sleep or hypnosis.

HYPNOSIS: An artificially induced sleep-like state.

HYPOACUSIA: Abnormally low auditory sensitivity.

HYPOAFFECTIVITY: Abnormally low general sensitivity. *Compare*: Hypoesthesia.

HYPOCALCEMIA: Abnormally low level of calcium in the blood. Hypocalcemia frequently occurs around the time of parturition, primarily in cattle (periparturient hypocalcemia). *Compare*: Milk Fever.

HYPOEMOTIVITY: Abnormally low emotional reactivity.

HYPOERGASIA: Abnormally low reactivity.

HYPOESTHESIA: Abnormally low sensitivity to stimulation. This term usually refers to specific sensory modalities (e.g., auditory, gustatory, olfactory, tactile, or visual hypoesthesia). *Compare*: Hypoaffectivity.

HYPOFERTILITY: Abnormally low reproductive fertility.

HYPOGALACTIA: Milk secretion by a postpartum female that is insufficient for raising offspring.

HYPOGLOSSAL NERVE: The twelfth cranial nerve which innervates the muscles of the tongue.

HYPOGLYCEMIA: Abnormally low blood glucose levels. Symptoms may include trembling, piloerection, depression, irritability, cold sweat, and hypothermia.

HYPOKINESIA: Abnormally low motor activity.

HYPOKINESIS: *See* Hypokinesia.

HYPOMAGNESEMIA: Abnormally low level of magnesium in the blood, often characterized by excessive eye and ear flicking and a tendency toward nervousness.

HYPOMETRIA: Motor action(s) insufficient to reach an object or goal.

HYPOMNESIA: Abnormally poor memory.

HYPOPHOSPHATEMIA: Abnormally low level of phosphorus in the blood, often characterized by sluggishness of movements and hip lameness.

HYPOPHYSIS: *See* Pituitary.

HYPOPNEA: Abnormally low pulmonary activity.

HYPOREFLEXIA: Abnormally low reflex activity.

HYPOSENSITIVITY: Abnormally low sensitivity to stimulation.

HYPOSEXUALITY: Abnormally low sexual excitability or generally low level of sexual activity.

HYPOSMIA: Abnormally low sensitivity to odors.

HYPOSTHENIA: Abnormal weakness.

HYPOSTIMULATION: Insufficient intensity or complexity of stimulation causing occupational vacuum and sensory deprivation.

HYPOSYNERGIA: Abnormally poor or distinctly defective coordination.

HYPOTHALAMUS: Part of the forebrain located below the thalamus. The hypothalamus controls maintenance of blood circulation, body temperature, urinary secretion, and influences related to hunger and thirst. It also secretes oxytocin and vasopressin.

HYPOTHERMIA: Abnormally low body temperature.

HYPOTHESIS, OBJECT: *See* Object Hypothesis.

HYPOTHETICAL CONSTRUCT: Any theoretical concept introduced with the aim of explaining the causation or function of behavioral responses.

HYPOTONIA: Abnormally low level of tonicity of skeletal muscles.

HYPOVENTILATION: Reduced flow of air in the lungs resulting in increased carbon dioxide levels in the blood.

HYPOXIA: Abnormally low supply of oxygen in the blood preventing normal functioning of body organs.

HYSTERESIS: A delay of effect or response after circumstances are changed by some stimulus or process. Hysteresis has been proposed to operate in situations of motivational competition to prevent inefficient behavioral strategies when the strengths of motivation for alternative actions are almost equal. For example, hysteresis would keep an animal that is equally hungry and thirsty from alternating excessively between food and water because the perceived change in its motivational state due to eating (or drinking) would be delayed for some time after the activity had begun.

HYSTERIA: A state characterized by extensive use of defense mechanisms and by a variety of symptoms associated with high levels of fear, anxiety, restlessness, and general irritability.

HYSTERIA, FARROWING: *See* Farrowing Hysteria.

I

ICTUS: A seizure, stroke, or any other sudden attack of disease.

IDEAL OBSERVER THEORY: A theory that only someone who has all relevant information and is free of biases could reliably distinguish what action is morally right or wrong. Inspired by this theory, some people claim that neither the farmers who raise animals for profit nor those welfarists who financially or otherwise benefit from their animal welfare related activities are unbiased in their approach to the issue of animal well-being.

IDEATION: The process of forming abstractions or reflecting memory images without their physical presence at the time. The term ideation is often used as distinct from processes involved in forming object hypotheses.

IDENTICAL TWINS: Twins of monozygotic origin.

IDIOPATHIC: Refers to a pathologic state of unknown cause. This term is also used in cases of behavioral disorders of unknown origin.

IDLING: Being stationary for extended periods of time while apparently not engaged in any other activity, such as rumination or sleep. Idling occurs predominantly in environments where an animal has relatively limited opportunity or stimulation to perform diverse behavioral activities and may be difficult to distinguish from resting, except that it is performed for more time than is actually required for resting.

ILLUSION: An object hypothesis that distinctly misrepresents reality. Illusions may involve different sense modes (e.g., optical, motion, temperature, taste, or auditory illusions). *Also see* specific illusion—OPTICAL; MOTION; VISUAL.

IMAGE, SEARCH: *See* Search Image.

IMITATIVE BEHAVIOR: Behavior of an organism that mimics that of other organisms.

IMMOBILITY: Lack of movement. *Also see* specific immobility—TONIC.

IMMOBILIZE: To render incapable of movement using physical restraint or an immobilizing agent, e.g., etorphine.

IMPERCEPTIBLE STIMULUS: A stimulus that is outside the threshold range of conscious perception.

IMPLANTS, GROWTH: *See* Growth Implants.

IMPLICIT RESPONSE: Any response not directly observable without the use of appropriate instruments.

IMPRESSION: In a psychological sense, the effect of sensory stimulation on the central nervous system prior to perceptual analysis.

IMPRINTING: Rapid learning to identify, approach, and follow some object (typically a parent). Imprinting occurs in some animal species during a critical period of development early in life and is strongly genetically predisposed.

IMPRINTING, ERRONEOUS: *See* Erroneous Imprinting.

INADEQUACY, MATERNAL: *See* Maternal Inadequacy.

INADEQUATE STIMULUS: A stimulus that activates sensory receptors considered to be inappropriate for such a stimulus (e.g., tactile contact with a soft object may stimulate thermal receptors and thus result in a perception of warmth). *Compare*: Adequate Stimulus.

INAPPETENCE: Failure to manifest the appetitive and consummatory phases of a behavioral cycle. The term generally is used in reference to food.

INATTENTION: Lack of attention or a state of unselective attention.

INBORN: Present in the organism at birth. Synonym: Innate.

INBREEDING: Sexual reproduction in which gametes originate from animals more closely related genetically than the average relationship in the population.

IN CALF (cattle): Referring to a pregnant cow. (colloquial term)

INCENTIVE: Anything that arouses or strengthens motivation for goal-directed behavior.

INCIDENTAL LEARNING: Learning that occurs without a detectable learning process or without ascertainable motivation.

INCIDENTAL REINFORCEMENT: Reinforcement that occurs unpredictably and rarely and is unplanned. Also called accidental reinforcement or spurious reinforcement.

INCIDENTAL STIMULUS: A stimulus that is not considered an essential part of a given experimental situation, but influences the behavior of the organism being studied.

INCITEMENT: A phenomenon in which an organism causes another to instigate an attack on a third organism. For example a female in the proximity of a male may show a threat display toward an approaching conspecific female, which may lead to the male driving the intruder away. Incitement generally is most common in monogamous species. *Compare*: Interference.

INCLUSIVE FITNESS: *See* Genetic Fitness.

INCOMMEASURABLE: Referring to two or more characteristics that cannot be classified with the same scale or unit (e.g., amount of feed consumed and speed of locomotion).

INCOMPLETE COITUS: *See* Coitus Incomplete, Coitus Interruptus.

INCOORDINATION: Insufficient synchronization of motor activity of muscles causing incoherent locomotion and reduced body balance.

INCUBATION: The maintenance of eggs in an environment that fosters embryonic development.

INDECISION: Inability to choose a course of action or select between alternatives.

INDEPENDENT VARIABLE: A variable that has the effect of inducing a change in another variable. An independent variable is generally controlled by the experimenter in order to determine its effect on a dependent variable.

INDEX, LANDAU: *See* Landau Index.

INDIFFERENCE POINT: The transition point between two opposing sensations (e.g., between cold and warm sensations).

INDIFFERENT STIMULUS: *See* Neutral Stimulus.

INDIVIDUAL DISTANCE: The distance an individual seeks to maintain between itself and other individuals.

INDIVIDUAL DISTANCE ZONE: An area surrounding an individual defined by the distance the individual seeks to maintain between itself and other individuals. The size of this zone depends on circumstances existing at a given time, e.g., reproductive stage, social status, temperature. This term generally is used in the context of conspecific social interactions. *Compare*: Group Distance Zone.

INDIVIDUALITY: A set of characteristics or qualities that distinguishes any given organism from its peers.

INDUCED MOLTING: *See* Controlled Molting.

INDUCED OVULATION: Ovulation that is dependent on a preceding event of copulation or artificial stimulation of the vagina and cervix. Rabbits, mink, and cats exhibit induced ovulation. *Compare*: Spontaneous Ovulation.

INERTIA: A tendency to persist in a state of rest or equilibrium. With regard to the nervous system this term denotes the time lag between a stimulus and the onset of a detectable response.

INFANTICIDE: Killing of young by a conspecific adult. *Compare*: Infantophagia.

INFANTILE BEHAVIOR: Behavior of neonatal animals during the stage when they are fully dependent on parental care.

INFANTILISM: Stalled or regressed psychophysiological development of an organism to the degree that it psychologically and behaviorally approximates an infant.

INFANTOPHAGIA: Eating of young by a conspecific adult. *Compare* Infanticide.

INFECTIOUS BRONCHITIS (poultry): Infectious inflammation of the bronchial tubes caused by *coronavirus*. Behavioral symptoms include coughing and sneezing at night. The infection results in reduced egg production and higher frequency of cracked eggs, soft shell eggs, and rough shell eggs.

INFERTILITY: Temporary loss of fertility. *Compare*: Sterility.

INFESTATION: Occurrence of large numbers of parasites within or on an organism.

INFLATION REFLEX: A respiratory reflex controlling the inspiration of air into the respiratory tract.

I

INFLUENZA: *See* specific influenza—EQUINE; SWINE.

IN FOAL (horse): Referring to a pregnant mare (colloquial term).

INFORMATION: That which is perceived by an organism. Any aspect of a stimulus that can be detected by the sensory modalities and accounted for by the organism.

INGESTIVE BEHAVIOR: Actions by which an organism takes substances into the body by swallowing.

INHALATION: Inspiration of air together with an airborn substance such as medicine. Antonym: Exhalation.

INHERENT VALUE: A value equal to all moral agents (I. Kant) or all subjects-of-a-life (T. Regan). Individuals within these categories are considered valuable in themselves, regardless of social status or utility to others. *Compare:* Intrinsic Value.

INHERITANCE: Acquisition of traits and propensities through the additive and nonadditive effects of genes contributed by each parent at conception.

INHIBITION: Partial or complete suppression of a response in the presence of an eliciting stimulus. *Also see* specific inhibition—AFFERENT; ASSOCIATIVE; CONDITIONED; EXTERNAL; MEMORY; REACTIVE; REINFORCEMENT; RETROACTIVE.

INITIATOR: An individual who incites imitative responses by other members of its group.

INJURY, SPREADER: *See* Spreader Injury.

IN MILK: Lactating. (colloquial term)

INNATE: *See* Inborn.

INNATE DRIVE: *See* Primary Drive.

INNATE RELEASING MECHANISM (IRM): A hypothetical mechanism selectively sensitive to releaser stimuli for a particular instinctive action. For example, initiation of suckling by neonates would be controlled by an IRM when contact is made with mammary teats.

INNER EAR: *See* Ear.

INNERVATION: The distribution of nerve axons on an organ.

INSEMINATION: Transfer of semen into the reproductive tract of a female. *Also see* specific insemination—ARTIFICIAL; NATURAL.

INSIGHT: A term used to refer to an ability of an organism to apprehend relationships among perceived or ideated events.

INSIGHT LEARNING: Learning characterized by apprehension of the features and possibilities of a novel situation so that the development of responses is not based solely on trial and error experience.

INSOMNIA: Extended inability to sleep.

INSPIRATION: A segment of the respiration cycle, characterized by expansion of the chest and lungs causing air intake.

INSTIGATOR: In a social context, an organism that initiates some action with regard to another organism (e.g., an animal instigating a fight).

INSTINCT: An unlearned tendency to act in ways typical of a species.

INSTINCTIVE BEHAVIOR: Any response that does not require a learning experience (frequently used synonymously with innate or inherited behavior).

INSTRUCTION: The process of systematically imparting messages.

INSTRUMENTAL CONDITIONING: A type of conditioning in which the organism must perform a specific response to obtain reinforcement. Methodologically, instrumental conditioning may be differentiated from operant conditioning in that each trial ends when the correct response is performed, whereupon the response-contingent reinforcement is presented.

INSTRUMENTAL ERROR: A constant measurement error caused by imprecise instrumentation.

INSTRUMENTAL VALUE: A quality that serves as a mean to generate something which is intrinsically valuable. *Compare*: Intrinsic Value.

INTEGUMENTARY: Pertaining to the skin or its layers and accessory structures such as hair, horns, hooves, and skin glands.

INTEGUMENTARY BEHAVIOR: Behavior directed toward the integument of the organism (e.g., grooming).

INTELLIGENCE: Ability of an organism to learn to cope with new situations and deal effectively with its environment.

INTENSITY: The quantitative aspect of a stimulus, sensation, or response.

INTENTION: Cognitive determination to perform a given action.

INTENTIONAL BEHAVIOR: Behavior directed toward a goal.

INTENTION MOVEMENT: An action that indicates the behavior in which an animal is about to engage. Frequently, such movements are merely preparatory to the subsequent behavior, or they may be the initial actions themselves. In conflict situations, an animal may display intention movements but not follow through with the behavior the intention movements suggest. In such cases, the movements may indicate which motivations are in conflict.

INTERACTION, SOCIAL: *See* Social Interaction.

INTERESTS: A term used in the formulation of theories of value that refers to positive (attractive) or negative (aversive) attitudes of a conscious organism toward surrounding objects or conditions of its own life. The objects or conditions of its own life that this organism likes or dislikes are considered to be a matter of its interest. Some animal welfare philosophers (e.g., P. Singer) define what is valuable to animals by referring to their interests. *Also see* Equal Consideration of Interests.

INTERFERENCE: A phenomenon in which an organism interrupts an aggressive encounter (threat or fight) among other organisms. Interference is most common in gregarious species. *Compare*: Incitement.

INTERIM ACTIVITIES: Behavioral actions that are induced by application of an intermittent reinforcement schedule and are displayed during the period between intervals of reinforcement.

INTERIOR: Situated inside or in the direction of the center.

INTERMITTENT REINFORCEMENT SCHEDULE: Any reinforcement schedule in which the reinforcing stimulus may not be presented upon performance of a correct response. With the exception of a response to reinforcement ratio of 1:1, all variable and fixed interval, and all variable and fixed ratio reinforcement schedules are intermittent reinforcement schedules. *Compare*: Continuous Reinforcement Schedule.

INTERNAL ENVIRONMENT: The total internal processes and states occurring within an organism's body.

INTEROCEPTIVE CONDITIONING: Classical conditioning applied to an internal physiological process using some stimulus in the internal environment as the conditioned stimulus.

INTEROCEPTIVE NERVOUS SYSTEM: Part of the peripheral nervous system that transmits afferent impulses from the viscera.

INTEROCEPTOR: A sensory receptor activated by stimuli occurring inside the body (e.g., proprioceptors and visceroceptors). Antonym: Exteroceptors.

INTERSEXUAL SELECTION: *See* Sexual Selection.

INTERSPECIES RELATIONSHIP: Typical forms of interaction between species, principally categorized as antagonism, amensalism, parasitism or parasitoidism, neutralism, commensalism, and mutualism or symbiosis.

INTERSTITIAL CELL STIMULATING HORMONE: A hormone of the anterior lobe of the pituitary gland that stimulates androgen secretion by interstitial cells of the testes. Synonym: Luteinizing Hormone.

INTERVAL, UNCERTAINTY: *See* Uncertainty Interval.

INTERVAL SCALE: A data scale in which recorded variables are categorized into mutually exclusive groups of equal class intervals from an unknown or arbitrarily set zero point (e.g., change in latency to perform some behavior by an experimental group of animals with the latency of the control group as the zero point). *Compare*: Ratio Scale.

INTOXICATION: A condition produced by excessive use of or application of toxic materials leading to behavioral distortion.

INTRASEXUAL SELECTION: *See* Sexual Selection.

INTRAVENOUS FEEDING: Artificially induced delivery of nutrients into the blood stream.

INTRINSIC BEHAVIOR: Behavior that is mediated apparently by specific response mechanisms and is performed with minimal plasticity (e.g., eye blinking).

INTRINSIC VALUE: Being valuable in itself. Something has intrinsic value if it is valued for itself rather than for what it leads to or what can be done with it. According to some animal welfare theories (e.g., that proposed by P. Singer), pleasures derived from satisfaction of preferences are intrinsically valuable to animals. *Compare:* Instrumental Value.

INTROMISSION: Insertion or admission into something. In a biological context, the term is most commonly used in reference to insertion of the penis into the vagina.

INTROSPECTIVE OBSERVATIONS: Observations of phenomena occurring within the body of the observer and directly detectable only by the observer himself. (e.g., pain, hunger, fear). Antonym: Extrospective Observation.

INTUITIONISM: A theory that judgements, including moral judgements, are known to be true without apparent processes of reasoning.

INVESTIGATIVE BEHAVIOR: Behavior of an organism indicative of inspection of an object or surroundings.

INVESTMENT, PARENTAL: *See* Parental Investment.

INVOLUNTARY BEHAVIOR: Any action that occurs without intention or volition (e.g., attention to an unexpected object).

INVOLUNTARY NERVOUS SYSTEM: *See* Autonomic Nervous System.

IRIS: A disc-shaped, pigmented membrane with a circular opening in its center (pupil) located in the anterior portion of the eye in front of the lens. By contracting or dilating, the iris can change the size of the pupil and control the amount of light entering the eye.

IRM: *See* Innate Releasing Mechanism.

IRREVERSIBILITY: Complete resistance of learned responses to extinction. *Compare*: Spontaneous Recovery.

IRRITABILITY: Ability to respond to stimulation. This term is used frequently to denote oversensitivity to stimulation.

ISHIHARA TEST: One of many tests to detect color blindness. It uses images of different hues that cannot be distinguished from the images' background by color-blind individuals.

ISOLATION: *See* Social Isolation.

ITCH: An irritating sensation that arouses motivation to scratch, rub, or bite the affected area of the body. A clinical level of itching may result from conditions caused by external parasites (e.g., mange, lice) or internal disorders (scrapie, Aujeszky's disease).

J

JENNY: A female ass.

JERKING, HEAD: *See* Head Jerking.

JIBBLING (horse): To balk at or defer a required action.

JOINTS, HYPERMOBILITY: *See* Hypermobility of Joints.

JUMP: To spring from the ground or surface with the propulsive force being derived primarily from the action of the legs.

JUMPER (horse): A horse trained, exhibited, or competing in jumping events. Jumpers are judged according to faults, such as knock-downs, touches, falls, refusals, run-outs, loss of gait or circling, and time penalties.

JUNCTION: *See* specific junction—MYONEURAL; NEUROMUSCULAR.

JUSTIFICATION, MORAL: *See* Moral Justification.

JUST-NOTICEABLE DIFFERENCE: The minimum difference between two intensities of stimulus that can be detected by an organism.

JUVENILE TEAT SUCKING: Habitual teat sucking occurring among juvenile, group-housed animals. Juvenile teat sucking can cause inflammation, abcesses, or stricture of the teat canals and subsequent inability to eject milk after parturition.

K

KAINISM: Killing and eating of sibs. *Compare*: Fratricide.

KAIROMONES: Biochemical substances emitted by an individual of one species that an organism of another species is able to detect and respond to (e.g., scent produced by an animal can be detected by flies and attract them to the animal). Kairomones are transmitted by air.

KASPAR-HAUSER ANIMAL: An animal that manifests behavioral abnormalities because of having been kept isolated for an extended period of time in an environment of very low complexity.

KEY STIMULUS: A stimulus that, in a given situation, plays a predominant role in elicitation of behavior or behavioral sequences.

KICKING: The act of delivering a powerful blow with one or both hind legs (e.g., by a horse). Habitual kicking is considered a dangerous vice, especially if directed toward humans or other animals. *Also see* specific kicking—STALL.

KID: A young goat up to one year of age.

KIDDING: Parturition in goats.

KIN: Any pair or group of individuals having relatively close genetic relationship through common ancestry.

KINE-: A prefix indicating a relationship to motion (e.g., kinesiology, kinesthesis).

KINESIOLOGY: Study of different types of muscles, muscle-leverage systems, and muscle movement.

KINESIS: Movement.

KINESTHESIS: Sensation of movement or strain in muscles, tendons, and joints.

KINESTHETIC: Pertaining to that part of the proprioceptive sense mode related to bodily movement.

KINETIC: Pertaining to motion.

KINKY-BACK (poultry): Postural adjustment in chickens caused by deformity of the vertebral column. (colloquial term)

KINOPSIS: A phenomenon in which individuals of a group are attracted to other group members by visual perception of their movement.

KIN SELECTION: Selection resulting from the effect that an organism has on the propagation of genes which are identical by descent to those carried by the organism itself but carried by kin (other than offspring).

KLINOKINESIS: Movement in changing direction.

KLINOTAXIS: Locomotion during which orientation is attained by regularly alternating lateral movements of a part of the whole body and comparing successive levels of stimulation intensities generated by directive factors.

KLÜVER-BUCY SYNDROME: A syndrome occurring as a consequence of bilateral temporal lobectomy, characterized by distinct alterations of behavior, such as increased sexual activity, sexual responsiveness to a variety of objects, hyperphagia, memory deficiency, and loss of fear.

KNOCKING, STALL: *See* Stall Kicking.

KNOWLEDGE: Any set of information available to an organism at any given time.

KRONISM: *See* Filicide.

K-SELECTION: Natural selection that favors high parental investment in individual offspring. High parental investment, in turn, limits the number of offspring that can be raised simultaneously. *Compare*: r-Selection.

KURTOSIS: A deviation from the shape of the normal frequency distribution curve, characterized by significant shortening or lengthening of the curve.

KYPHOSIS: Abnormal ventroflexion of the spine. Kyphosis may be a symptom of abdominal pain, rabies, or tetanus.

LABOR: A period of time during which the fetus(es) is expelled from the body of the mother. Labor may be divided into distinct stages: dilatation, expulsion, and placental. The dilatation stage begins with the onset of regular uterine contractions and ends when the birth canal is completely prepared for passage of the fetus. The expulsion stage starts when the fetus begins to move through the birth canal and ends when it is completely expelled from the mother's body. The placental stage begins at the end of the expulsion stage and ends when the placenta and membranes have been expelled.

LABORING ACTION (horse): Locomotion performed with or requiring excessive effort. *Compare*: Free Going.

LABYRINTH: An intricate combination of passages between an entrance and exit (e.g., the cavities and canals constituting the internal ear, or an apparatus to study orientation and spatial memorization).

LACRIMAL REFLEX: Increased production of tears in response to corneal irritation.

LACTATION: The act of secretion of milk or the period of secretion of milk.

LACTOGENIC: Stimulating milk secretion.

LACTOGENIC HORMONE: *See* Luteotropin.

LAG: *See* Response Latency.

LAMB: A young sheep up to one year of age.

LAMBING: Parturition in sheep.

LAMENESS: Deviation from normal gaits caused by injury or disease (e.g., foot-and-mouth disease, laminitis, milk lameness, etc.). Lameness can be further subdivided into supportive or suspensory lameness, and sudden, progressive or chronic lameness. *Also see* specific lameness—CHRONIC; PROGRESSIVE; SUDDEN; SUPPORTIVE; SUSPENSORY.

LAMINITIS: Inflammation of laminae below the outer horny wall of the hoof generally characterized by reluctance to move, increased time spent lying down, and body postures indicating intention to reduce weight pressure on the affected leg(s).

LANDAU INDEX: A procedure for quantifying a social hierarchy, calculated as the degree of deviation from a linear dominance order.

LANDMARKS: Permanently located cues aiding orientation (e.g., tree, building, etc.).

LANGUAGE, DANCE: *See* Dance Language.

LATENCY: *See* specific latency—REFLEX; REPRESENTATION; RESPONSE.

LATENCY TIME: *See* Response Latency.

LATENT: Hidden or not immediately apparent.

LATENT LEARNING: Learning that is not apparent and occurs in the absence of detectable reinforcement. Latent learning may promote an animal's ability to learn in the future. For example animals that have an opportunity to investigate an area or apparatus prior to learning trials subsequently perform better in the same area or apparatus than animals not given such an opportunity.

LATENT TIME: *See* Response Latency.

LATERAL: Pertaining to a position to the side of the median part of a body.

LATERAL GAIT: A gait in which the lateral pairs of legs move at the same time (e.g., pace).

LATERAL RECUMBENCY: Lying characterized by full lateral contact of the body trunk, neck, head, and legs on one side with the ground. The neck, head, and legs are usually extended during lateral recumbency. The frequency of lateral recumbency varies among different species of farm animals and different age groups and may also be influenced by environmental circumstances (e.g., type of ground surface, ambient temperature, time of day).

LAW: An established rule. It may refer to general principles of the way things work in nature (natural law), to morality of human actions (moral law), or enforced social agreements (civil law). *Also see* specific law—YERKES-DODSON.

LAW OF ANALOGY: A generalization suggested by E. Thorndike stating that organisms respond to a new situation as they have initially responded to similar situations in the past.

LAW OF HETEROGENEOUS SUMMATION: A generalization stating that independent components of a compound stimulus have additive effects on the behavioral response of an organism.

LAYING, EGG: *See* Oviposition.

LD CYCLE: A cycle with alternating light and dark periods. LD 16:8 designates 16 hours of light alternating with 8 hours of darkness. LL and DD designate continuous illumination or continuous darkness, respectively.

LD-50: A dose of substance that is lethal to the median (50%) of all organisms of a given species or type. The LD-50 test has been severely criticized for causing considerable suffering of animals without necessarily providing useful information.

LEADERSHIP: The ability of an individual to control or direct the behavior of the members of a group.

LEADING LEG: The foreleg in a canter or gallop that is not diagonally synchronized with one of the hind legs and lands last in the gait sequence. In an animal traversing a circle, the leading leg is typically that on the inside of the circle.

LEANING: Partial support of the head or body on some object such as a wall, fence, or animal, achieved by shifting the center of gravity horizontally from a point of balance over the legs.

LEARNED BEHAVIOR: Any action performed as a result of, or influenced by, experience. This does not mean that learning is independent of inheritance, since the ability to collect, memorize, and utilize information for the learning process is influenced by genotype.

LEARNED DRIVE: *See* Secondary Drive.

LEARNED HELPLESSNESS: A phenomenon that may occur when avoidance or reduction of some aversive stimulation is impossible or extremely difficult for the subject. Learned helplessness is manifested by continued acceptance of such stimulation even when the situation subsequently provides opportunity for an avoidance response.

LEARNED PREFERENCE: Any preference influenced by a conditioning process.

LEARNING: The ontogenic process of attainment and memorization of information. *Also see* specific learning—INCIDENTAL; INSIGHT; LATENT; MOTOR; OBSERVATIONAL; PERCEPTUAL; PREPROGRAMMED.

LEARNING CURVE: Graphic illustration of changes in an animal's responsiveness to the learning process, arranged according to consecutive units of practice.

LEARNING DISPOSITION: A theoretical concept referring to propensity of an organism to acquire associations in a given situation. The learning disposition of an organism is determined by genetic makeup, psychological and physical condition, and prior learning.

LEARNING METHOD, COMPLETE: *See* Complete Learning Method.

LEARNING SET: Mental representation of a principle common to a series of learning trials such that, although circumstances may differ between trials, the solution to each is governed by the common principle. An animal is considered to have acquired a learning set when it apprehends the principle.

LEE-BOOT EFFECT: Suppression of initiation of estrous behavior by overcrowding of animals.

LEG DROP (horse): An action in which a horse rapidly tucks a leg under its body and lowers itself somewhat toward the quarter with the tucked leg.

The leg drop is performed to avoid bites directed at the legs in aggressive social encounters.

LEG, LEADING: *See* Leading Leg.

LEG WEAKNESS SYNDROME: A clinical term for a variety of defects in the structure and function of the legs (e.g., broad hips, narrow lumbar area, hypermobility of joints, abnormal angulation of the foot, etc.). Behavioral symptoms are difficult rising from recumbency, staggering gait, inability to maintain positional balance, extended period of recumbency, and paresis.

LEG YIELDING (horse): A dressage maneuver in which the axis of the horse's body is maintained at about 30 degrees from the direction of movement. The head of the horse is slightly bent at the poll so that it points even farther from the direction of movement.

LEK: A small breeding territory characteristic of some species of birds and mammals (e.g., wild pheasant).

LENS: A transparent structure located in the anterior portion of the eye that focuses rays of light down onto the retina.

LESION: Severe damage or loss of tissue characterized by discontinuity of the tissue's function.

LETDOWN: *See* Milk Letdown.

LETHAL: Causing death.

LETHARGY: A state of pronounced drowsiness or loss of desire to act.

LETHISIMULATION: Complete temporary immobilization of an animal, visually approximating its death, induced by a sudden and inescapable exposure to a strong fear causing stimulus.

LEUCOSIS: A complex of viral diseases occurring mainly in fowl. Behavioral symptoms are progressive hypoactivity, termination of egg laying, ruffled feathering, paralysis, and death.

LEVADE (horse): A dressage exercise of the Spanish High Riding School in which the horse raises the anterior portion of its body (approx. 30 degrees) with its hind legs spread apart and bent at the hocks, and the forelegs folded underneath the chest.

LIBERTARIANISM: A theory guided by the principle that rightness or wrongness of human actions depends on whether such actions promote or restrict the liberties of human beings. In relation to the treatment of farm animals, libertarians tend to emphasize that only the owner of the animals is responsible for the treatment of his/her animals.

LIBIDO: Sexual motivation. The level of libido often is assessed from the reaction time between first detection of a sexual partner and first mounting attempt by the male or display of receptivity posture by the female.

LICKING: An action whereby the tongue is slid over the surface of an object. Licking may have a nutritional, gustatory, prophylactic, or social function and also may develop as a stereotyped behavior pattern. Licking consists of self-licking (an animal licks its own body) or social licking (an animal licks another animal), also called allolicking. A special category of social licking is sexual licking which occurs predominantly during courtship and estrus. *Also see* specific licking—SEXUAL; SOCIAL.

LICKING SYNDROME: Excessive occurrence of social licking, assumed to be an indication of sodium deficiency.

LIE: To assume or maintain a recumbent position. The behavior patterns involved in lying down show a high degree of uniformity within species, but may be modified by environmental circumstances (e.g., restriction of movement, injury). The posture when lying can be categorized as lateral recumbency, semi-lateral recumbency, or ventral or sternal recumbency. The body position of a recumbent animal may reflect health status (e.g., unusual posture to prevent pressure on an injured or inflamed area), level of exhaustion (e.g., selectiveness of place to lie and tendency to adopt lateral recumbency), degree of thermal comfort (e.g., the area of exposed body surface), and level of relaxation (e.g., position of the head, eyelids, or limbs).

LIFE, QUALITY OF: *See* Quality of Life.

LIFE, SUBJECT OF A: *See* Subject-of-a-Life.

LIFT, REAR-LEG (horse): *See* Rear-leg Lift.

LIGHT: Radiant energy in the range of frequencies that stimulates visual receptors.

LIGHTING, SUPPLEMENTAL: *See* Supplemental Lighting.

LIGHT REFLEX: Contraction of the pupil when exposed to light.

LIGNOPHAGIA: The eating of wood or wooden substances. Lignophagia is commonly identified as wood gnawing.

LIMEN: *See* Threshold.

LIMINAL SENSITIVITY: The lowest level of stimulation that evokes a detectable response.

LIMINAL STIMULUS: A stimulus that is close to the detection threshold of an organism and thus barely evokes a response.

LIMITED FEEDING: Any feeding program that provides less feed than the animals would consume *ad libitum.* Such a program frequently is used to prevent excessive fattening of animals. Synonym: Restricted Feeding.

LINE: A group of animals interrelated through an ancestor(s) and possessing some common genetic and phenotypic characteristics.

LINEAR HIERARCHY: A type of social hierarchy formed as a straight ranking line of dominance-subordinance relationships through the whole group. The highest ranking animal dominates all, the second highest all but the first, the third highest all but the first two, etc. down to the animal subordinate to all.

LINEAR-TENDING HIERARCHY: A type of social hierarchy in which dominance-subordinance ranking includes a single hierarchy loop. *Compare*: Complex Hierarchy.

LINEBREEDING: Sexual reproduction in which gametes originate from animals related to a common ancestor(s). Linebreeding can be considered a special form of inbreeding to cumulate the genes of such an ancestor(s).

LINECROSSING: Sexual reproduction in which gametes originate from animals belonging to two separate genetic lines.

LIP-CURLING: A response to pheromones, manifested by inhalation followed by distinct upward curling of the upper lip and dorsiflexion of the neck. Lip-curling is most noticeable in male ungulates when in proximity to estrous females or when examining their genitalia, urine, or urine marking but may also be exhibited by female bovines when close to estrous peers. Synonym: Flehmen.

LIPOSTATIC THEORY: A theory that proposes that hunger and satiation reflect an organism's endeavour to maintain a constant blood lipid level.

LIP SMACKING (horse): A rhythmical smacking action of the lower lip toward the upper lip. Lip smacking is exhibited by some horses when under saddle or in unfamiliar environments. Lip smacking is thought to be indicative of mild excitement and is considered to be a nuisance vice.

LISTEN: To attend to sound stimuli.

LISTERIOSIS: An infectious disease, caused by *Listeria monocytogenes*, which may be manifested in the form of meningoencephalitis (characterized by circling, swallowing difficulty, blindness, head pressing, perpetual chewing motion, and abortion occurring mainly in the second half of pregnancy) and septicemia (affecting mainly young animals and characterized by weakness, depression, occasionally severe diarrhea, and liver necrosis). Listeriosis occurs both in ruminants and monogastric animals.

LITTER: A group of siblings born during one parturition period. *Also see* specific litter—DEEP.

LIVESTOCK: Farm animals used for production purposes.

LOBE: A naturally existing, partially separated portion of an organ (e.g., olfactory lobe of the brain, superior lobe of the lungs, hepatic lobes, or pituitary gland lobe).

LOCALITY MEMORY: The ability of an organism to identify particular areas.

LOCALIZED STIMULUS: A stimulus applied to a particular area of the body or organ.

LOCK, COITAL: *See* Coital Lock.

LOCK-JAW: *See* Tetanus.

LOCOMOTION: The activity of an organism moving itself between points in space. *Also see* specific locomotion—DIGITIGRADE; PLANTIGRADE; RICOCHETAL.

LOCOMOTOR: An organ that has potential to facilitate locomotion.

LOCUS, GENE: *See* Gene Locus.

LOLLING, TONGUE: *See* Tongue Lolling.

LONG-TERM MEMORY: Memory that is retained for extended periods and is relatively resistant to disruption or loss. Long-term memory is characterized by a relatively large capacity for information storage and is thought to have a different control system than short-term memory.

LOOSE HOUSING: A housing system where animals are not restricted in ambulation within a pen.

LORDOSIS: Dorsiflexion of the spine occurring during the stage of sexual receptivity of females of some mammalian species.

LOWER CRITICAL TEMPERATURE: *See* Critical Temperature.

LOW SET: A term used to designate a short-legged animal.

LUMBO-SACRAL PLEXUS: A network of nerves in the hind limb connected to the last three lumbar and the first one (pigs, dogs) or two (horses, cattle, sheep) sacral nerves of the spinal cord.

LUMINOSITY: The quality of radiating or reflecting light.

LUNGING (horse): The process of training a horse using a long rein attached to the noseband of a bitless bridle.

LUTEAL PHASE: The stage of the reproductive cycle in which there is an active corpus luteum and a high secretion of progesterone.

LUTEINIZING HORMONE (LH): A hormone of the anterior lobe of the pituitary gland that stimulates ovulation and luteinization in females. Synonym: Interstitial Cell Stimulating Hormone.

LUTEOTROPIN: A hormone of the anterior lobe of the pituitary gland that maintains the corpus luteum and stimulates lactation. This hormone is also called lactogenic hormone or prolactin.

LYING: Maintaining a recumbent position.

M

MACROSMATIC: Having a highly developed olfactory perceptual ability or sense of smell. Antonym: Microsmatic.

MAD COW DISEASE: *See* Bovine Spongiform Encephalopathy.

MAIDEN (horse): A horse that has never won a race; a mare that has never been bred.

MAINTENANCE BEHAVIOR: Any behavior through which an organism sustains its own physiological equilibrium or that of a dependent. In a broader sense, this term refers to activities required for essential physical and psychological comfort.

MALADAPTATION: A state of some behavioral or other inadequacy that causes the genetic contribution of an organism to the gene pool of the subsequent generation to be less than the average of its peers.

MALADAPTIVE BEHAVIOR: Any behavior that directly or indirectly causes reduction or inhibition of the reproductive performance of a given organism. *Compare:* Maladjusted Behavior.

MALADJUSTED BEHAVIOR: Any behavior that directly or indirectly reduces production performance, leads to health problems, or causes discomfort of a given organism. *Compare:* Maladaptive Behavior.

MALADJUSTMENT: Failure of an organism to cope effectively with its surroundings.

MALE, SOMNOLENT: *See* Somnolent Male.

MALFORMATION: Structural abnormality in the development of bodily organs.

MALNUTRITION: A state of extended inadequate nutrition caused by unbalanced diet or defective metabolism.

MALSUCKING: Sucking directed toward other objects than mammary teats of a lactating female, or sucking directed toward the mammary teats but performed at an inappropriate age. Malsucking of inappropriate objects occurs frequently in early weaned or young, bucket-fed calves. Malsucking among adult animals is considered a serious social vice.

MAMMILLA, OR MAMILLA: The nipple of a mammary gland.

MANGE: A parasitic disease of farm animals caused by infestation of various types of mites. Behavioral symptoms are rubbing, scratching, or shaking of affected parts of the body. *Also see* specific mange—SARCOPTIC.

MANIA: A general term for behavioral disorders indicative of disease of the nervous system. Characteristic actions are excessive licking, chewing of unusual materials, perceptual incompetence, staggering gait, abnormal vocal-ization, and increased aggressiveness.

MANIPULANDUM: *See* Operandum.

MAP, COGNITIVE: *See* Cognitive Map.

MARE: A female horse that has foaled, or that is older than four years.

MAREK'S DISEASE: *See* Leucosis.

MARKER ANIMAL: A hormonally masculinized female or a vasectomized or penectomized male used for identification of estrous cows.

MARKER, CHIN: *See* Chin Marker.

MARKING: Any deposition of scent compounds or other signs that serve to de-lineate territorial boundaries or trails, or to designate partners and group members.

MASCULINITY: The condition of having well developed secondary male sex characteristics and displaying typical male behavior.

MASCULINIZATION: Acquisition of male behavioral or body conformational characteristics by a female.

MASKING (horse): Defecation or urination, generally by a stallion, on the fe-ces of another horse. Masking is postulated to be an assertion of status or territory, or an attempt to mask olfactory stimuli from the feces of the other horse.

MASSAGING: Rhythmic and sustained mechanical manipulation of the integu-ment and underlying tissues (e.g., manipulation of the mammaries of a sow by piglets prior to and after nursing).

MASTICATION: The process of physically breaking down food in the mouth by tearing and grinding action of the teeth in preparation for swallowing and digestion.

MASTITIS: Inflammation of the mammary glands or udder that occurs in all mammals, but most commonly in cows and ewes. Symptoms in cows are signs of pain, particularly when the affected part of the udder experiences pressure, and reduced milk yield, with the milk often having an abnormal appearance. There may be swelling and hardening of the infected region.

MASTURBATION: Sexual stimulation of the genitalia without coitus. Masturbation has been recorded in males of all mammalian farm animal species. Masturbation is conducted by rhythmic movements approximating pelvic thrusting (horses, cattle) or by self-licking of the penis (dogs, cats).

MATE: An organism that is the sexual partner of an individual of the opposite sex. Such an association may be of short duration or may involve a pair bond over an extended period of time. As a verb, the term means to form a sexual partnership or, more specifically, to copulate.

MATE PREFERENCE: A phenomenon manifested by some males and females of many polygamous and polyandrous species to favor specific sexual partners while avoiding or reducing interactions with others. Mate selection has been observed in horses, sheep, goats, and pigs.

MATERNAL BEHAVIOR: Behavior of a dam directed toward the well-being and care of her dependent offspring or young.

MATERNAL DEPARTURE: The phenomenon in which a dam temporarily leaves her offspring in some place (e.g., a nest), enabling her to conduct activities essential for survival (e.g., foraging for food).

MATERNAL INADEQUACY: Inability of a dam to provide sufficient maternal care to her neonatal offspring. Maternal inadequacy usually is caused by agalactia or hypogalactia, undernourishment, poor health, excessive excitability, or exhaustion due to dystocia. *Compare:* Maternal Rejection.

MATERNAL REJECTION: Social rejection of neonatal offspring by their dam. Maternal rejection develops when, for some reason, normal primary socialization does not take place. It occurs in all species of farm animals, most frequently with primiparous dams or dams having infected or injured udders. *Compare:* Maternal Inadequacy, Kronism.

MATING: *See* specific mating—ASSORTATIVE; RANDOM.

MATING BEHAVIOR: The component of sexual behavior closely associated with and including copulation.

MATING STANCE: Posture of a female indicative of sexual receptivity. Most common characteristics of this stance in four-legged animals are slightly spread hind legs, extended neck with slightly lowered head, and laterally displaced tail.

MATRIARCH: A female that is predominant in the social organization of a social unit. Such predominance involves alpha social status and arises out of ancestry to other members of the unit.

MATURATION: The process of development through infant and juvenile stages to adulthood.

MATURITY, SEXUAL: *See* Sexual Maturity.

MAVERICK: A farm animal not marked with an owner's brand.

MAXIMAL SENSATION: An intensity of sensation that cannot be increased by further stimulation.

MAZE: An experimental apparatus, consisting of a combination of pathways, used to study learning and motivation.

MEASUREMENT: *See* specific measurement—BEHAVIORAL; DIRECT.

MEATUS: The tubular passage of the external ear extending from the auricle to the tympanic membrane. The meatus facilitates transmission of sound waves to the middle ear.

MECHANICAL STIMULATION: Activation of receptors by means of pressure, vibration, or movement.

MECHANISM: A hypothetical unit of the total psychophysiological system involved in activation, manifestation, or inhibition of a particular behavior. This term often is used in association with other words (e.g., response mechanism, behavioral mechanism, releasing mechanism, innate releasing mechanism, and feedback mechanism). *Also see* specific mechanism—FILTERING.

MECHANISTIC EXPLANATION: Interpretation of all causes of observable processes only by physical and chemical means. Mechanistic explanations ignore the role of abstractions and disregard concepts such as desires, goals, or purposes.

MECHANOCEPTOR: A sensory receptor activated by any mechanical stimulation occurring outside or within the body of the organism.

MECONIUM: Fetal feces that accumulate in the colon during the course of intrauterine development and are discharged soon after birth.

MEDIAN: A plane or point of division of some distribution into two equal parts.

MEDICAMENTOUS CASTING: Rendering an organism temporarily immobile through administration of chemical compounds.

MEDULLA OBLONGATA: The bulb at the top of the spinal cord. The medulla oblongata is the location of nerve centers controlling autonomic functions such as respiration, heart rate, and gastrointestinal processes.

MEISSNER CORPUSCLES: Structures in the skin containing tactile receptors.

MEMBRANE, TYMPANIC: *See* Tympanic Membrane.

MEMORY: The entire set of information acquired and retained by an organism and potentially available for its use. *Also see* specific memory—LONG-TERM; SHORT-TERM.

MEMORY INHIBITION: *See* Memory Interference.

MEMORY INTERFERENCE: Disruption of consolidation of memory information due to another learning experience. Interference may be proactive (consolidation is hindered because the animal has learned something else already) or retroactive (consolidation is hindered because the organism subsequently learns something else). The interference tends to be greatest when the interfering information is similar to the memorized information. Memory interference differs from extinction or habituation. Synonyms of proactive and retroactive interference are proactive and retroactive inhibition, respectively.

MENDELISM: Explanation of inheritance based on transmission of genes from parents to offspring. The two basic principles of Mendelian inheritance are segregation and independent assortment of genes.

MENINGES: Membranes (dura mater, pia mater, and arachnoid) that envelop the brain and spinal cord.

MENINGITIS: Inflammation of the membranes covering the brain or spinal cord. Behavioral symptoms are restlessness, excitability, erratic locomotion, increased respiration and increased vocalization followed by hypo-activity, retention of urine and feces, occasional rolling over and circling, frequent head resting, and progressive paralysis.

MENTAL: Pertaining to the cognitive state or activity of the brain.

MESENCEPHALON: *See* Midbrain.

METABOLISM: Complex physiological processes encompassing the digestion of nutrients, formation of bodily substances, and release of energy for bodily functions ranging from simple biochemical reactions to complex behavioral patterns.

METESTRUS: The postovulatory stage of the estrous cycle, characterized by formation of corpora lutea.

METHOD, MINIMAL CHANGE: *See* Minimal Change Method.

METHODOLOGY: Study of the principles of scientific inquiry.

MICROSMATIC: Having a poorly developed olfactory perceptual ability. Antonym: Macrosmatic.

MICTION: *See* Urination.

MICTURITION: *See* Urination.

MICTURITION REFLEX: The sequence of activities controlling urination. The reflex center is located in the spinal cord but is also controlled by the brain.

MIDBRAIN: The middle part of the vertebrate brain, consisting of the two cerebral peduncles and the tectum. The midbrain develops from the central part (mesencephalon) of the three divisions of the embryonic brain.

MIDDLE EAR: *See* Ear.

MIGRATION: Movement, often seasonal and distant, of part or all of an animal population from one area to another.

MIGRATION EXCITEMENT: Excitement associated with migration or preparation for migration.

MILK: A liquid secretion from the mammary glands of post-partum female mammals. Milk is the primary source of nutrients for preweaned mammalian young, and is composed principally of water, lactose, fat, casein and other proteins, minerals, and vitamins. As a verb, the term means to express milk, generally by a human or milking machine, from the mammary glands of some animal, e.g., a dairy cow. Also see specific milk—BEE; RESIDUAL; UTERINE; WITCH'S.

MILK EJECTION: A process, controlled by the hormone oxytocin, characterized by contraction of myoepithelial cells, alveoles, and small ducts of the mammary gland resulting in the expulsion of secreted milk.

MILK FEVER: A syndrome occurring in high milk-yield dairy cows and goats caused by hypocalcemia. Symptoms include initial hyperexcitability, dilation of the pupils, and cool extremities, followed by staggering gait, difficulty in rising, and extended recumbency, which can lead to muscle damage and paralysis of the hind limbs. Fever is not a clinical feature of this disease.

MILKING: Manual or mechanical extraction of milk from the mammary gland.

MILKING HOBBLES (cattle): A simple device used to prevent kicking. It consists of two metal pieces shaped to fit the hindlegs above the hocks and connected in front by a chain. At the present time, milking hobbles are seldom used during milking but occasionally are employed for the safety of a person conducting medical examination or treatment.

MILKING PARLOR: An area specially designed for milking.

MILK LETDOWN: Milk ejection. (colloquial term)

MIMESIS: Manifestation of imitative behavior.

MINCING: Circling around an adversary with the body hunched up and taking short, rapid steps. This behavior appears to be a combination of an attacking and of a defensive posture, commonly interpreted as a threat display.

MINIMAL CHANGE METHOD: Any technique designed to detect the minimal threshold differentiation ability of an organism. The procedure involves exposing the organism to a series of stimuli that ascend or descend from a standard stimulus. Occasionally called just-noticeable difference method.

MISE EN MAIN (horse): Relaxation of the jaw during ramener in dressage.

MISIMPRINTING: *See* Erroneous Imprinting.

"MM" VOCALIZATION (cattle): A low-amplitude sound of variable duration (0.5-3.0 sec) produced with the mouth closed or only slightly open. It is emitted during social interactions with offspring and familiar peers (particularly young animals), during sexual interactions between females, or (directed toward a caretaker) while waiting for food or milking. It is thought that this vocalization promotes maintenance of herd cohesion.

MOB: To engage as a group in harassment or attack of an individual(s).

MODAL ACTION PATTERN: Any action pattern typical of a given species or breed. Modal action pattern was coined as an alternative to fixed action pattern in recognition that such action patterns vary somewhat between individuals and with circumstances.

MODALITY, SENSE: *See* Sense Mode.

MODE: In a statistical context, the value with the highest frequency on a distribution scale. *Also see* specific mode—SENSE.

MODIFICATION, BEHAVIORAL: *See* Behavioral Modification.

MOLAR CATEGORY: Any category that subsumes entities having a broad range of characteristics.

MOLTING: Seasonal shedding and regrowth of the pelage of mammals and the plumage of birds. *Also see* specific molting—CONTROLLED.

MOLTING, CONTROLLED: *See* Controlled Molting.

MONDAY MORNING DISEASE: *See* Exertional Myopathy.

MONOCULAR VISION: Vision in which images are projected on the retina of one eye only.

MONOESTROUS FEMALES: Females that manifest only one estrous cycle per year.

MONOGAMY: A reproductory arrangement in which two individuals of opposite sex form a primary and durable breeding unit.

MONOGASTRIC: Referring to an animal having a single, uncompartmented stomach.

MONOPLEGIA: Paralysis or paresis of one limb only.

MONORCHID: A male that possesses only one testicle in the scrotum.

MONOSYNAPTIC REFLEX: A reflex in which the afferent neurons synapse directly with efferent neurons without intermediate neurons being a part of the reflex arc. *Compare:* Polysynaptic Reflex.

MONOTOCOUS: Typically delivering only one offspring at each parturition (e.g., cattle, horses). Antonym: Polytocous.

MONSTERS: Neonatal animals born with morphological deformities. (colloquial term)

"MOOEH" VOCALIZATION (cattle): A high amplitude call of intermediate duration (0.7-1.2 sec) usually produced in a series of two to fifteen calls with the neck extended and mouth wide open, generally by sexually mature males. This call is a threat signal when directed toward potential male adversaries at a distance or a greeting signal when oriented toward females or distant herdmates.

"MOOH" VOCALIZATION (cattle): A medium to high amplitude sound of intermediate duration (0.5-1.5 sec) produced as a single sound or a series of two to ten calls with neck fully extended and mouth open. It is emitted when an animal is thirsty or hungry, separated from its calf, or threatened by a perceived danger, and is thought to be an indication of distress.

MOON BLINDNESS (horse): Equine periodic opthalmia, an eye disease. At an early stage it occurs periodically (approximately as often as every new moon; thus, the name moon blindness). As the disease progresses, cataract formation leads to eventual blindness.

"MOO" VOCALIZATION (cattle): A medium amplitude sound of widely varying duration (0.5-4.0 sec) produced with neck partially extended and mouth open. It is emitted by animals seeking others or when hearing familiar sounds of feed delivery. It is thought that this call is an indicator of mild excitement.

MORAL AGENT: Someone who is capable of understanding moral norms and obligations and able to determine how to act in accordance with them. A being that is aware of its moral obligations. Rational human beings, in contrast to animals, are considered moral agents. *Compare:* Moral Patient.

MORAL EGOISM: *See* Rational Egoism.

MORAL JUSTIFICATION: Reasoning which proves that a statement or an action is morally acceptable. The recognition of animal ability to suffer calls for justification of animal treatment by humans. Animal welfarists demand that treatment of sentient animals ought to be morally acceptable in all stages of animal life.

MORALLY RELEVANT DIFFERENCE: For any two creatures 'd' is a morally relevant difference between them if by reference to 'd' we can explain why we have some moral obligation to one of the creatures which we do not have to the other. According to some animal rights views, humans are morally obligated not to kill animals for food because there is no morally relevant difference between people and animals; that is, given that it is morally wrong to kill people for food, if there is no difference 'd' between people and animals, then it is morally wrong to kill animals for food.

MORAL PATIENT: A being who has interests and whose interests ought to be taken into account by moral agents, but who does not have a moral obligation itself. A moral patient lacks the capacity to act according to moral obligations. Animals, in contrast to rational human beings, are considered moral patients. *Compare:* Moral Agent.

MORAL RELATIVISM: A theory guided by the principle that there is no universal moral code valid for all human societies and cultures; the moral acceptability of an action is thus determined by specific cultural or societal norms. In relation to animals, moral relativists tend to tolerate differences between culturally identifiable societies in their treatment of animals.

MORAL THEORIES, CONSEQUENTIALIST: *See* Consequentialist Moral Theories.

MORAL THEORIES, DEONTOLOGICAL: *See* Deontological Moral Theories.

MORBIDITY: A state of sickness. Also refers to the proportion of animals in a herd or flock affected by a given disease.

MORIBUND BEHAVIOR: Behavior an animal performs when it is dying.

MORPHOLOGY: The study of forms and structures of living organisms.

MORTALITY: Termination of life. Numerical expressions of mortality are useful only if related to age (e.g., perinatal, neonatal, prepubertal mortality) or to components of the population at a given time.

MORTAR-EATING: Licking or pecking surfaces rich in minerals. It is assumed to be a sign of mineral deficiency in the diet.

MOTHER: *See* Dam.

MOTHER, FOSTER: *See* Allomother.

MOTHERING: *See* Maternal Care. (colloquial term)

MOTHERING ABILITY: The ability of a dam to provide maternal care to her offspring. Various assessments of mothering ability commonly are used as criteria in breeding programs involving selection of female agricultural mammals.

MOTILE: Capable of locomotion.

MOTION ILLUSION: An illusion that an object is moving when it actually is motionless.

MOTIONLESS SITTING: Maintaining a dog-sitting position for a long period of time with lowered head, semi-closed eyes, and apparently low level of alertness. The behavior occurs most often in pigs housed individually on concrete floors and fed concentrated diets.

MOTIVATION: Urge to perform a given behavioral action. Motivation arises when a neural controlling mechanism that interprets incoming stimuli and assesses available alternatives selects an appropriate response. *Compare:* Drive.

MOTIVATIONAL ANALYSIS: An analysis that focusses on the detection of psychophysiological causes of a given behavior.

MOTIVATIONAL COMPETITION: A phenomenon wherein motivation for an alternate behavioral action becomes preeminent to that for an ongoing behavioral action, and the animal changes its behavior. In a situation of motivational competition, the motivation for the behavioral action that is supplanted does not decrease prior to the change in behavior. *Compare:* Disinhibition.

MOTIVATIONAL CORNERING: Cornering induced by some strong social or psychological attachment, and not because of physical restraint. (For example, a female in the presence of her offspring may not flee a predator, even if she could, but may instead attack it.) *Compare:* Cornering.

MOTOR: Referring to structures and functions of muscular movement and glandular processes.

MOTOR ACTIVITY: Any activity resulting from the excitation of muscles and glands.

MOTOR AREA: An area in the frontal lobe of the cerebrum that elicits activity of skeletal muscles when it is stimulated.

MOTOR LEARNING: Learning that is described in terms of the activity of muscles and glands.

MOTOR NERVOUS SYSTEM: Part of the somatic nervous system consisting of efferent nerves that innervate skeletal muscles.

MOTOR NEURON: An efferent neuron connected to a muscle or gland. Through excitatory and inhibitory impulses, motor neurons control the activity of such organs.

MOTOR UNIT: A group of muscle fibers innervated by the branches of one neuron.

MOUNT: An event of mounting.

MOUNTING: The act of an animal raising the anterior part of its body generally onto the posterior part of another animal. Mounting, while typically performed by a male as part of copulatory behavior, is also common in sexually homogeneous groups of animals. Mounting among females is usually a reliable sign of estrus. *Also see* Mounting Misalignment.

MOUNTING MISALIGNMENT: A heterosexual attempt to mount a female in a way that will not facilitate coital intromission of the male's penis; e.g., head mounting or flank mounting.

M

MOURNING: A state of psychological depression, caused by separation from strongly bonded organisms or objects, or by pain and illness. According to circumstances, behavioral indicators may include distress calls, lowered mobility, reduced appetite, and weakened attention to occurrences in its surrounding.

MOUTH: The cavity in the head delineated by the jaws and pallet and containing the teeth and tongue. In birds the mouth includes the beak. At the anterior, the mouth opens to the external environment through the oral orifice (also called the mouth), and at the posterior connects with the nasal cavities, esophagus, and trachea. The mouth is adapted for the intake, gustation, mastication, and ingestion of food, facilitates vocalization and tactile stimulation, and may be used for breathing when the nasal passages are blocked or large volumes of air are required. In some species, the mouth also may serve as a weapon. *Also see* specific mouth—BROKEN.

MOUTHING: Investigation of objects by manipulation with the mouth. Mouthing is most common in young animals. In some situations, mouthing may develop into a persistent, stereotyped behavior pattern.

MOVEMENT: Any spatial translocation of all or part of an organism. *Also see* specific movement—COMFORT; EXPRESSIVE; FETAL; INTENTION; SIGNAL; STEREOTYPED.

"MRRH" VOCALIZATION (cattle): A variable amplitude sound of a relatively short duration (0.5-1.0 sec) produced with the mouth closed. It occurs when animals are resting, but may indicate muscular effort in a resting situation because it frequently is emitted when an animal adjusts its lying posture. In mature males, a similar sound often is emitted in conjunction with propulsus during copulation.

MULE: The offspring of an ass sire (jack) and a horse dam.

MULEY: A polled cow.

MULTICHOICE APPARATUS: A device constructed to offer a choice among three or more alternatives.

MULTIPAROUS: Referring to females having undergone more than one cycle of pregnancy and parturition. *Compare:* Primiparous.

MULTIPLE REINFORCEMENT SCHEDULE: An operant type of reinforcement schedule that uses various reinforcement schedules depending on the occurrence of discriminative cues. For example, bar pressing to receive a reward may be reinforced by a fixed interval schedule in the presence of one type of light, but reinforced by a fixed ratio schedule in the presence of a different type of light.

MUSCLE FATIGUE: Inability or severely reduced ability of a muscle fiber to exert force when action potentials pass through the neuromuscular junction and spread over the fiber. Muscle fatigue develops as a consequence of prolonged and strong contraction of the muscle involved.

MUSCLING, DOUBLE: *See* Double Muscling.

MUSCONE (swine): A pheromonal steroid produced in the sub-maxillary gland of mature boars and secreted through the prepuce. A synthetic form of this steroid is used as an estrus detection aid in swine.

MUSCULATURE: The complete muscular system of a body or of any of its parts.

MUSTANG: A feral horse descended from horses brought by Spanish colonists to America.

MUTATION: Alteration of the nucleic acid sequence in the DNA complement of a gene, or change in chromosome structure. Mutations in the DNA complement of a germ cell may be heritable.

MUTILATION: The process of cutting, tearing, or otherwise removing normal, healthy parts of an organism's body.

MUTUALISM: *See* Symbiosis.

MYOCLONUS: The rapid, repeated and involuntary contraction and relaxation of muscles (muscle spasm).

MYOGLOBINURIA: Reddish or dark-brown discoloration of the urine due to the presence of myoglobin. This condition occurs in association with exertional myopathy.

MYOGRAM: Graphic representation of muscular activity (contraction) produced by a myograph.

MYOLOGY: The study of muscles.

MYONEURAL JUNCTION: *See* Neuromuscular Junction.

MYOSITIS: Inflammation of muscles that may result from injuries or physical overload. Behavioral symptoms are avoidance of muscular tension in affected areas, swelling, hardening, and sensitivity to pressure.

MYOTATIC REFLEX: Reflexive contraction of a muscle when it is stretched. Also called stretch reflex.

N

NAIVE: Lacking in experience.

NANISM: Dwarfism.

NAPPING (horse): Serious disobedience of the rider's commands. Napping may be manifested in various forms, such as refusing to leave the home range territory, refusing to jump, or ignoring direction commands.

NARCOLEPSY: A condition in which the affected animal may suddenly and uncontrollably fall asleep, particularly when excited. Narcolepsy has been observed in some breeds of horses (e.g., Suffolk horses, Shetland ponies) and in dogs and humans.

NARCOSIS: A state of stupor induced by chemicals.

NASAL CATARRH: *See* Rhinitis.

NASAL REFLEX: Sneezing in response to irritation of the nasal membrane.

NATIVISM: A theory that emphasizes that all functions of organisms are innate.

NATURAL: Pertaining to nature, accepted the way things are, spontaneous, innate, primitive, etc. In an ethological sense, any behavior or situation that develops and occurs without human intervention. A term with a very broad spectrum of meanings and therefore to be used with caution.

NATURAL BREEDING: Breeding that takes place without direct human intervention.

NATURAL INSEMINATION: Insemination achieved by copulation.

NATURALISM: A belief that mental phenomena are determined only by neurophysiological processes in the brain and, as such, are subject to evolutionary changes.

NATURALISTIC OBSERVATIONS: *See* Field Behavioral Observations.

NATURAL LAW THEORY: A theory that claims that there are moral laws of nature parallel to the physical laws of nature. According to this theory, the moral laws can be discovered by human reason. Creatures lacking reason could neither discern nor regulate their behavior in accordance with moral laws.

NATURAL MATING: Natural insemination.

NATURAL PACER: An animal that performs the pace gait without training by humans. Camels and some horses are natural pacers.

NATURAL PREFERENCE: Any preference that is not influenced by a learning experience. Synonym: Instinctive Preference.

NATURAL SELECTION: *See* Selection.

NATURAL SERVICE: Natural insemination.

NAUSEA: An unpleasant sensation that may culminate in vomiting.

NAVEL SUCKING: Sucking on the navel area of other individuals. It occurs often in young, early-weaned, group-housed, trough or bucket-fed calves. Navel sucking can cause serious damage to the affected tissue, resulting in infection and abscesses.

NEAR SIDE (horse): The left side of the horse.

NECK, ARCHED (horse): *See* Arched Neck.

NECK EXTENDING: Pronounced forward extension of the neck manifested by many animals, but most distinctly by geese. Neck extending in geese seems to have two meanings: a greeting signal when accompanied by gabbling, or a warning signal when accompanied by hissing.

NECROSIS: Death of a portion of a tissue or organ. *Also see* specific necrosis— CEREBROCORTICAL.

NECROSIS, CEREBROCORTICAL: *See* Cerebrocortical Necrosis.

NEED: Any requirement that is necessary for an organism to develop normally and to maintain its physical and psychological health. *Compare:* Desire.

NEEDLE TEETH (swine): A pair of temporary tusks and a pair of corner incisors in the upper and the lower jaws of neonatal piglets.

NEGATIVE FEEDBACK: A process in which the performance of an activity leads to reduction or termination of the activity. Eating, for example, has a negative feedback on appetite, which reduces further motivation to eat.

NEGATIVE REINFORCER: *See* Reinforcer.

NEIGH (horse): A medium to high amplitude, rapidly pulsed sound of relatively long duration (1.0-3.0 sec). This sound is produced with the mouth partially open and nostrils slightly vibrating. The head usually is raised and the ears point forward. It is emitted when a horse is separated from individuals with which it has a social bond (e.g., a dam separated from offspring, a horse separated from a permanent stablemate or sexual partner during estrus). It is thought to be an indication of excitement.

NEONATE: A newborn organism.

NEONURSING: Nursing that occurs during and shortly after parturition, characterized by sporadic and socially unsynchronized suckling by neonatal piglets prior to establishment of a regular nursing cycle.

NEOPHOBIA: The fear of novel stimuli. Many organisms exhibit neophobia, which is commonly induced by unfamiliar or altered surroundings or unusual events.

NEOTENY: Appearance of juvenile morphological or behavioral characteristics in adults. In agricultural species of animals, neoteny may occur as a result of selective breeding to produce animals that are more manageable as adults (e.g., lack of horns in species whose adults typically have horns).

NEPHRITIS: Inflammation of the kidneys. Behavioral symptoms may include stiffness, arched back, reduced appetite, and discharge of small amounts of blood-stained urine.

NEPOTISM: Behavior in favor of individuals that are genetically related to the one manifesting the behavior.

NERVE: A bundle of nerve fibers held together by connective tissue. *Also see* specific nerve—ABDUCENS; ACCESSORY; ACOUSTIC; AUDITORY; FACIAL; GLOS-SOPHARYNGEAL; HYPOGLOSSAL; OCULOMOTOR; OLFACTORY; OPTIC; TRIGEMI-NAL; TROCHLEAR; VAGUS.

NERVE BLOCK: Conduction Anesthesia. (colloquial term)

NERVE IMPULSE: *See* Action Potential.

NERVOUS SYSTEM: The entire array of neurons in the body of an organism, responsive to internal and external stimulation. Neurons, interconnected by synapses, form a network that penetrates the whole body and facilitates rapid transmission of information between sensory receptors and effectors along pathways that range from very simple and direct (for reflexes) to very complex (for problem solving or decision making). In higher organisms, the nervous system is composed of two subsystems called the central nervous system (or cerebrospinal system) and the peripheral nervous system. *Also see* specific nervous system—AUTONOMIC; CENTRAL; EXTEROCEPTIVE; INTE-ROCEPTIVE; INVOLUNTARY; MOTOR; PARASYMPATHETIC; PERIPHERAL; PROPRIO-CEPTIVE; SOMATIC; SOMESTHETIC; SYMPATHETIC.

NEST: A structure built by some animals for the purpose of depositing and hatching their eggs, delivering and protecting their offspring, or in some cases, housing themselves.

NESTING: The process of building or occupying a nest.

NEURAL ARC: *See* Neural Circuit.

NEURAL CIRCUIT: A functional unit mediating conduction of action potentials from receptors through connecting neurons to effectors. Neural circuit is also called neural arc or sensorimotor arc.

NEURAL CONDUCTION: Transmission of action potentials along a nerve fiber and from neuron to neuron.

NEURITIS: Inflammation of nerves, followed by paralysis of those part(s) of the body innervated by the affected fibers.

NEUROETHOLOGY: A scientific study of neural control of behavioral processes and activities.

NEUROLOGY: Study of the structures and functions of the nervous system.

NEUROMUSCULAR JUNCTION: The point of contact between the axonal branch of a motor neuron and a muscle fiber. Except for the multi-inner-vated muscle fibers of poultry, each muscle fiber has only one neuromuscu-lar junction. The action potential of the nerve axon is transmitted to the muscle fiber by the release of "quanta" or packets of acetycholine that are associated with synaptic vesicles. A new action potential is initiated on the muscle fiber. Acetylcholine is rapidly broken down by acetylcholinesterase in the narrow gap between the axonal membrane and the muscle fiber mem-brane.

NEURON: A basic cell of the nervous system specialized to carry action potentials. A typical neuron has a cell body with a nucleus and two or more fibrous extensions called axons and dendrites.

NEUROSIS: A mild psychological disorder usually caused by unresolved conflicts or anxiety. *Also see* specific neurosis—EXPERIMENTAL.

NEUROTIC: Referring to an individual whose behavior is indicative of a minor nervous disorder.

NEUROTOXIN: Any substance harmful to nerve tissue (e.g., botulin).

NEUROTRANSDUCER: A neuron able to synthesize compounds that act as a functional link between the nervous system and a gland.

NEUROTRANSMITTER: Any substance that facilitates transmission of action potentials between nerve terminals (e.g., acetylcholine, epinephrine, or norepinephrine).

NEUTRALISM: A form of interspecies relationship in which neither species affects the other.

NEUTRAL OBJECT: *See* Neutral Stimulus.

NEUTRAL STIMULUS: A stimulus that does not cause conditioning if applied independently of an unconditioned stimulus.

NEWCASTLE DISEASE (poultry): An infectious viral disease of fowl. Symptoms include depressed egg production, higher occurrence of soft shell eggs, occurrence of diarrhea, and twitching of the head and twisting of the neck.

NIBBLING, WITHER: *See* Wither Nibbling.

NICHE: An ecological concept referring to the total range of environmental features occupied, and the amount of each used by the individuals of a given species. The niche is determined both by environmental opportunity and by the behavioral and physical attributes of the species in question.

NICKER (horse): A low to medium amplitude sound of relatively short duration (0.4-1.0 sec). This sound is produced with the mouth closed and nostrils dilated by horses expecting food, by dams toward offspring (quiet nicker), and by stallions during courtship (forceful nicker).

NICKING: Deliberate surgical severing of any tendon or muscle in the tail of a horse.

NIGHT BLINDNESS: *See* Nyctalopia.

NOCIRECEPTOR: Any sensory receptor that facilitates the perception of pain.

NOCTURNAL: Pertaining to the night. *Compare:* Crepuscular, Diurnal.

NOCTURNAL ORGANISM: An organism that performs most of the behavioral actions in its repertoire at night and rests predominantly during the hours of daylight.

NOESIS: The mental process involved in reasoning or judgment.

NOMINAL SCALE: A data scale in which recorded variables are categorized into mutually exclusive unranked groups (e.g., responding, nonresponding; lying, standing still, walking). *Compare:* Ordinal Scale.

NONADAPTIVE BEHAVIOR: *See* Maladaptive Behavior.

NONPARAMETRIC STATISTICS: A branch of statistics that makes no assumption regarding the distribution characteristics of data.

NONRETURN RATE: The proportion of first inseminations, or first postpartum inseminations, that result in pregnancy, as related to a given sire, herd, or service region. With artificial insemination in cattle, this is traditionally calculated as a 60-90 day nonreturn rate. This technical parameter is considered an important animal reproduction variable.

NONRUMINANT: Animals without a rumen as part of their digestive tract (e.g., pigs and poultry).

NONSEASONAL BREEDING: Breeding that in natural circumstances occurs during the whole year, although not necessarily with the same intensity (e.g., cattle, swine, and chickens are nonseasonal breeders).

NONSWEATING SYNDROME: *See* Anhydrosis.

NORADRENALIN: *See* Epinephrine.

NOREPINEPHRINE: *See* Epinephrine.

NORMAL BEHAVIOR: Behavior that qualitatively and quantitatively does not deviate from regular or stabilized form. Commonly interpreted as behavior of an animal that is healthy and free from pathological stress.

NORMAL CURVE: *See* Normal Distribution Curve.

NORMAL DISTRIBUTION CURVE: A symmetrical, bell-shaped distribution curve having the mean, median, and mode equal to the same value and possessing statistically convenient probability characteristics. The standard normal curve has a mean of 0 and a variance of 1. Synonym: Normal Curve, Gaussian Curve.

NORMAL EMBRYONIC POSITION (poultry): Typical fetal position in hatching eggs conducive to successful hatching. This position is characterized by the long axis of the body being parallel to the long axis of the egg and the beak being located underneath the right wing with the top of the beak directed toward the air cell in the blunt end of the egg. *Compare:* Embryonic Malposition.

NORMATIVE PRINCIPLE: A rule considered to serve as a primary guideline to distinguish between right and wrong human actions. In a philosophical sense a normative principle derives from some ethical theory. With regard to the treatment of animals the normative principle may reflect animal rights, utilitarianism, or another applicable theory.

NOSE: A facial structure of mammals having two external openings (nostrils) that lead to nasal cavities which in turn connect with the buccal cavity and the respiratory passages. The nose is used as a pathway for transfer of air to and from the lungs during breathing, and the nasal cavities contain olfactory receptors which enable olfaction.

NOSE LEAD (cattle): A small device that can be inserted into an animal's nostrils and squeezed by hand to pinch the nasal septum and thus facilitate handling.

NOSE RING (cattle): A strong metal ring, approximately 8-9 cm in diameter, permanently inserted into the nostrils of a bull and penetrating the nasal septum. The nose ring is used to facilitate handling.

NOSING: Making contact with an object or the body of another organism with the nose. Nosing is commonly displayed during investigation following sniffing, prior to suckling, and during courtship.

NOSING CALLS (swine): High frequency sounds of irregular duration produced by piglets during the nosing stage of the nursing event.

NOTCHING: Cutting of notches in the ears of animals for the purpose of permanent identification.

NOXIOUS: Unpleasant, painful, harmful, or injurious.

NOXIOUS STIMULUS: A stimulus that is unpleasant or harmful.

NUDGING: Relatively gentle pushing generally using the head. Nudging is displayed by cattle, sheep, goats, and swine during play and most typically by males during sexual courtship or immediately before mounting the female.

NULLIPAROUS: Referring to females that have not given birth to viable offspring.

NUPTIAL FLIGHT: Joint flight of male(s) and female(s) preceding their mating (e.g., flight of a queen bee and drones).

NURSING: The act of releasing milk to suckling young. *Also see* specific nursing—CYCLICAL.

NURSING CYCLE: A series of nursing-related activities culminating in ejection of milk by the nursing dam and ingestion of such milk by suckling young. Swine exhibit a complex nursing cycle consisting of several phases: Assembly phase (gathering of piglets around the sow and assumption of the nursing posture by the sow); Nosing phase (massage of the mammary glands by the piglets, accompanied by nosing calls and occasional suckling attempts); Slow suckling phase (noningestive suckling conducted at a rate of approximately 1 pulse per second); True suckling phase (ingestive suckling conducted at a rate of approximately 3 pulses per second); and Departure phase (sporadic massaging, suckling, and occasional resting, followed by complete separation of piglets from the mammary glands).

NURSING GRUNT (swine): One of an extended rhythmic series of grunts produced by a sow just prior to and during the nursing cycle. The rate of grunting changes from slow (approx. 1 per sec) before milk letdown to fast (approx. 3 per sec) during milk letdown and then gradually decreases when milk letdown ceases. Nursing grunts alert piglets to the imminent onset of milk availability and appear to help coordinate their suckling activity.

NURTURE: The total spectrum of environmental effects on an organism from the time of conception. Also, the provision of care by one individual to another individual or to itself.

NUTRIENTS: A variety of chemical compounds required by an organism for metabolic processes necessary for survival, growth, and good health. Nutrients consist of water, carbohydrates, lipids, proteins, minerals, and vitamins.

NUTRITIONAL WISDOM: Seeking of specific nutrients to compensate for deficiencies in the diet. In a broader sense, the preferential consumption of foods containing nutrients that the organism requires.

NYCTALOPIA: Inability or severely reduced ability to see in darkness. This visual disorder develops as a consequence of vitamin A deficiency.

NYMPHOMANIA: Exaggerated manifestation of sexual behavior in females.

NYSTAGMUS: Periodic, apparently involuntary vertical, horizontal, or rotatory eye movement. The eye movement usually proceeds slowly, but when the endpoint of the motion is reached, there is a quick return of the eye to its original orientation.

OBESITY: Excessive accumulation of body fat resulting in abnormally high body weight.

OBJECT: An observable physical entity.

OBJECT, NEUTRAL: *See* Neutral Stimulus.

OBJECT HYPOTHESIS: Subjective interpretation of objects or events. The hypothesis is incomplete or in error if the interpretation of such objects or events differs from reality. *Compare:* Perception, Illusion.

OBJECTIVE SCORE: Any measurement score that is independent of subjective assessment.

OBSERVATION: An examination process to ascertain certain facts or particulars about a reality.

OBSERVATION, DIRECT: *See* Direct Observation.

OBSERVATIONAL LEARNING: Learning by observing the behavioral responses and reinforcement of other individuals. Observational learning also is called vicarious learning.

OBSERVATIONS, EXPERIMENTAL BEHAVIORAL: *See* Experimental Behavioral Observations.

OBSERVATIONS, FIELD BEHAVIORAL: *See* Field Behavioral Observations.

OBSERVATIONS, NATURALISTIC: *See* Field Behavioral Observations.

OBSERVER: An individual attentive to ongoing phenomena. This term is most frequently applied to a person whose intention is to record and report on such phenomena.

OBSERVER RELIABILITY: The degree of agreement of a person's observation record with a known standard or another independently obtained observation record of identical behavioral occurrences. The agreement can be expressed in Pearson's "r" coefficient or in "percent agreement."

OBTUSION: Underdeveloped or otherwise permanently reduced sensation.

OCCUPATIONAL VACUUM: A state of seriously thwarted motivation of an organism to interact with its environment, caused either by diminished novelty or unarousing surroundings. Occupational vacuum can also be generated by inability of the organism to interact with its surroundings due to extended physical or psychological restraint.

OCELLUS: A simple invertebrate eye. Honeybees have three ocelli that function in directional orientation using the angle of rays from a light source, usually the sun.

OCULAR: Referring to the eye.

OCULOMOTOR NERVE: The third cranial nerve which innervates all eye muscles except the lateral rectus and superior oblique muscles.

ODDITY TRAINING: A series of discrimination trials in which the correct solution is characterized by being different from a group of alternatives. The learning set to be acquired involves the principle that the correct alternative to choose has features not shared by any other alternative.

ODOR: Sensation caused by chemical stimulation of receptors in the mucous membranes of the nasal cavities. *Also see* specific odor—GROUP.

OESTRUS: *See* Estrus.

OFFENSIVE AGGRESSION: Unprovoked aggression. In a social context, offensive aggression is not manifested in response to aggression instigated by another individual, but is an attempt to gain some resource at the expense of the individual toward which it is directed.

OFF FEED: Referring to loss of appetite for food. (colloquial term)

OFF SIDE (horse): *See* Far Side.

OFFSPRING PREFERENCE: Preference of a young animal to associate with and receive care from a particular adult(s). Such a preference may be ex-

O

pressed at more than one level, e.g., preference for a parent versus another adult, or for one parent versus the other.

OGIVE CURVE: An S-shaped curve depicting cumulative frequencies over the range of a normal distribution.

OILING: A component action of preening in which a bird uses its beak to spread oil secreted by the uropygial gland over its plummage. Oiling helps to keep the feathers in good condition.

OLFACT: A unit of liminal sensitivity in perception of odor, expressed as substance concentration per unit volume.

OLFACTION: The process whereby dispersed molecules of one or more substances are sensed as smells.

OLFACTORY: Referring or relating to the sense of smell.

OLFACTORY AREA: The area of the cerebral cortex where the center of olfactory perception is located.

OLFACTORY BULB: Part of the forebrain from which the olfactory nerve extends to olfactory receptors.

OLFACTORY CELLS: Spindle-shaped cells containing receptors sensitive to odors and located in the membranes of the nasal cavities.

OLFACTORY NERVE: The first cranial nerve which connects the olfactory receptors with the olfactory bulb in the brain.

OLIGURIA: Reduction in the amount of urine excretion. *Compare:* Anuria.

OMEGA ANIMAL: An animal that ranks lowest socially in its group. *Compare:* Alpha Animal.

OMNIVOROUS: Subsisting on feed of both plant and animal origin.

ONAGER: The wild ass native to southwestern Asia.

ONTOGENIC: Referring to the development of a particular individual. *Compare:* Phylogenic.

OOGENESIS: The process of formation of female gametes.

OPEN-FIELD BEHAVIOR: Behavior manifested and/or studied in an open-field test situation.

OPEN-FIELD TEST: Exposure of an organism to some standardized environmental arrangement, apart from that where it normally resides, in order to study its approach-avoidance responses or signs of emotionality.

OPERANDUM: Part of a conditioning apparatus designed to be operated by an organism to attain reinforcement (e.g., a bar in an automatic waterer to be pressed by cattle to attain water, or a disc to be pecked by birds to attain feed).

OPERANT: Referring to a behavioral action interpreted in the context of its immediate function or its effect on the environment.

OPERANT CONDITIONING: A type of conditioning in which reinforcement is contingent on a given operation (e.g., bar pressing). In operant conditioning it is unimportant what technique the animals use to conduct such an operation; the response may be somewhat variable as long as the operation is performed.

OPERATIONAL DEFINITION: Definition of a concept in terms of antecedent conditions, actions, and consequences. For example, the operational definition of hunger may deal with food deprivation time and subsequent action manifested by the deprived organism.

OPERATIONALISM: A theory that claims that all meaningful scientific terms either refer directly to observable phenomena or are definable through the use of operational definitions. *Compare:* Operational Definition.

OPTHALMIA: Contagious inflammation of the eye, affecting mainly sheep and goats.

OPTICAL ILLUSION: An illusion with regard to spatial relationships (e.g., the illusion that one line is longer than another of equal length when the former has obtuse angles at both ends while the latter has acute angles at both ends).

OPTIC CHIASMA: The area of the hypothalamus where the fibers of the optic nerves cross each other. Also called optic commissure or optic decussation.

OPTIC NERVE: The second cranial nerve which connects the retina with visual centers in the brain.

ORAL: Pertaining to the mouth.

ORAL TENDENCY: Tendency to examine objects using the mouth. Normally, oral tendencies are common in very young mammals. Artificially, they can be induced by bilateral temporal lobectomy.

ORDINAL SCALE: A data scale in which recorded variables are categorized into mutually exclusive groups arranged with reference to their magnitude (e.g., low-high, slow-intermediate-fast). *Compare:* Nominal Scale.

ORGAN: Part of the body that performs a special function or functions. *Also see* specific organ—ACCESSORY.

ORGANIC DISORDER: A disorder resulting from structural damage to an organ.

ORGANISM: A living, self-contained, integrated system of closely interdependent parts functioning together to carry out vital activities (e.g., an animal, plant, or bacterium).

ORGANIZATION, SOCIAL: *See* Social Organization.

ORGANOGENESIS: The process of development of bodily organs.

ORGASM: The culminating point of sexual excitement associated with distinct physiological and behavioral characteristics such as high vasopression, rhythmic muscular contractions, and semen ejaculation (in males).

ORIENTATION: A state of being or process of becoming aware of position in space.

ORIENTING RESPONSE: A response by which an animal gives selective at-

O

tention to a particular stimulus in its surroundings. It may be manifested as sensory focussing, body movement, etc.

ORIGINAL STIMULUS: In generalization studies, a conditioned stimulus to which other stimuli (test stimuli) are compared.

ORTHOKINESIS: Straight, or approximately straight, movement.

ORTS: The remains of feed in a trough or feeder that an animal(s) refrains from eating.

OSMORECEPTOR: A specialized cell in the hypothalamus that monitors the level of dehydration of an organism. Stimulation of osmoreceptors generally leads to increased effort to obtain water to drink and initiation of physiological strategies to conserve water.

OSSICLES, TYMPANIC: *See* Tympanic Ossicles.

OSTEITIS: Inflammation of the bone substance, characterized by signs of pain, lameness, postural and locomotory abnormalities, and progressively more time spent in a recumbent position.

OSTEO-: A prefix meaning "bone" (e.g., osteogenesis, osteology).

OSTEODYSTROPHY: Suppressed development or abnormal metabolism of bones, causing postural abnormalities, defective gait, and susceptibility to bone fracture.

OSTEOPOROSIS: A skeletal disease that is characterized by an atrophy of skeletal tissues, causing increased fragility and often a deformation of bones and ambulation difficulties.

OUTBREEDING: Sexual reproduction in which gametes originate from animals less closely related genetically than the average relationship in the population.

OUTCROSSING: Sexual reproduction in which gametes of one sex originate from animals subjected to inbreeding for several generations and gametes of the other sex originate from animals which are less related genetically than the average genetic relationship in the population. Outcrossing is a special case of outbreeding.

OUTDOOR HOUSING: A housing system where animals are kept predominantly or continuously outdoors. Outdoor housing facilities usually have simple shelters to provide protection against weather extremes.

OVARY: A female sexual gland producing ova and estrogen.

OVERCROWDING: An excessively high spatial density of animals due to some abnormal environment (e.g., inadequate space caused by physical constraint or avoidance of an inhospitable area). Overcrowding compels breakdown of individual distance zones and, in extreme circumstances, does not permit animals to rest at the same time, stand up or lie down freely, extend their limbs without interference, or have adequate opportunity for eating or drinking. The consequences of overcrowding depend on its intensity and duration, and thus may range from frustration, boredom, elevated competition for space, and lowered cleanliness in the pen to serious behavior anomalies (stereotypy, lethargy, learned helplessness), high levels of aggression, nutritional deprivation, inability to maintain effective thermoregulation, and increased mortality. *Compare:* Crowding.

OVERINCLUSION: Inability to inhibit an undesirable response associated with a particular stimulus.

OVERLEARNING: Continuation of learning trials beyond the stage necessary for required retention.

OVERLYING (swine): Incidental lying down by a sow onto piglets, causing their injury or death.

OVER REACH (horse): Defective leg action causing the hoof of the hind leg to touch the heel of the foreleg.

OVERSHADOWING: A phenomenon in which the reinforcing properties of a reinforcer are diminished by the concurrent presentation of some strong, distracting stimulus.

OVERSTIMULATION: *See* Hyperstimulation.

OVERTRAINING: *See* Overlearning.

OVERT RESPONSE: Any easily observable response.

OVINE: Sheep or pertaining to sheep.

OVIPAROUS: Reproducing by laying fertile eggs from which the offspring hatch outside the body of the mother. *Compare:* Viviparous.

OVIPOSITION: The act of expulsion of fully developed eggs from the body.

OVULATION: Discharge of a secondary oocyte from an ovarian follicle. *Also see* specific ovulation—INDUCED; SILENT; SPONTANEOUS.

OVUM: A female reproductive cell.

OX: A castrated male bovine used in draft work.

OXYTOCIN: A hormone of the posterior lobe of the pituitary gland that stimulates uterine contraction and milk ejection.

OZENA: *See* Rhinitis.

P

PACE (horse): A two-beat gait with the leg movement synchronized laterally. The sequence of hoof beats is left hind together with left fore followed by right hind together with right fore. Generally, the pace is an artificially attained gait in horses. *Compare:* Natural Pacer.

PACER (horse): A horse that races using a two-beat lateral gait called the pace. *Also see* specific pacer—NATURAL.

PACING: Stereotyped, short-distance walking back and forth, or side to side, typically manifested by animals kept in close confinement.

PACINI CORPUSCLES: Touch receptors located in the skin.

PADDLING (horse): Defective leg action in which the leg is thrown outward during the extension phase.

PAIN: Unpleasant sensation, usually localized, resulting from noxious stimulation or injury.

PAIN SYMPTOM: Any sign or behavioral display indicative of distinct discomfort due to the experience of pain, such as lack of or reduced eye movement, puckered eyelids, dilated nostrils, ears pulled back (horse), teeth grinding (swine), reduced or lack of tonguing (cattle), head turning and kicking toward affected part of the body, distress vocalization, abnormal postures, frequently changing body position, pawing with the feet, head pushing toward the wall, sensitivity to palpation of the affected area, etc.

PAIR BOND: *See* Bond.

PAIRING: Any voluntary or imposed grouping of two individuals. Typical pair formations can be observed between sexual partners, between mutually investigating adversaries, etc.

PALATABILITY: Relative acceptability of feed as influenced by factors such as taste, smell, structure, color, or other nonnutritional attributes.

PALATE: Sense of taste. Also, the roof of the oral cavity.

PALOMINO (horse): A golden colored horse with flaxen or silvery mane and tail.

PALPATION: Any touching conducted for the purpose of diagnosis. This term also is used occasionally to describe the action of a male's forelegs along a female's flanks during mounting.

PALPITATION: Exaggerated heart beat.

PALSY: Paralysis.

PANDICULATION: *See* Stretching.

PANPSYCHISM: A theory that all living and nonliving objects in the universe possess some psychological qualities more or less similar to humans and some level of awareness of other objects in their surroundings.

PANSEXUALISM: An attempt to explain the motivation of behavior in terms of sex drive.

PANTING: Increased breathing characterized by polypnea or hyperpnea and generally performed with an open mouth and distinct thoracic movements. Panting primarily serves to increase oxygen-carbon dioxide exchange or to increase evaporative heat loss.

PAPILLAE: Small protruberances containing receptors of touch, smell, or taste (e.g., papillae of the tongue possess receptors of taste).

PARADIGM, BEHAVIORAL: *See* Behavioral Paradigm.

PARADOXICAL SLEEP: *See* REM-Sleep.

PARALLEL MOVEMENT: Completely synchronized movement of dual organs. This term is most commonly applied to movement of the eyeballs when focussing on distant objects.

PARALYSIS: Partial or complete loss of muscle functions caused by injury or neural disfunction.

PARALYTIC MYOGLOBINURIA: *See* Exertional Myopathy.

PARAPLEGIA: Paralysis or paresis of the posterior pair of limbs.

PARASITE: Any organism practicing parasitism.

PARASITISM: A form of interspecies relationship in which individuals of one species benefit at the expense of individuals of another but, in so doing, do not kill them.

PARASITOIDISM: A form of interspecies relationship in which individuals of one species benefit at the expense of individuals of another species and, in so doing, cause their death.

PARASYMPATHETIC NERVOUS SYSTEM: Part of the autonomic nervous system branching out from the cranial and sacral portions of the central nervous system. Parasympathetic stimulation lowers heart rate and blood pressure, constricts the pupils and bronchi, and increases activity of the digestive tract. *Compare:* Sympathetic Nervous System.

PARENTAL: Pertaining to parents, to adults displaying parental care, or to the generation of parents as a whole.

PARENTAL BEHAVIOR: Behavior of parents, foster parents, or other adults characterized by assistance, cooperation, leadership or training of offspring or young conspecifics prior to weaning.

PARENTAL INVESTMENT: Expenditure of resources or performance of behavior on behalf of offspring at the expense of the ability to do the same for other offspring.

PARENTAL PREFERENCE: Preference of a parent to care for or associate with a particular young individual or group of such individuals. Such a preference may be expressed at several levels, e.g., preference for offspring versus other young, for neonatal offspring versus older offspring, or for certain individuals versus others in a litter.

PARENTAL REJECTION: Social rejection of offspring by a parent.

PARESIS: Partial paralysis.

PARESIS, PARTURIENT: *See* Milk Fever.

PARLOR, MILKING: *See* Milking Parlor.

PARTIAL CORRELATION: Correlation between two variables adjusted for the effect of other variables.

PARTIALLY ANTERIOR PRESENTATION: Fetal presentation in which the nose and one foreleg enter the birth canal while the other foreleg is retained below the fetus' body.

PARTURIENT PARESIS: *See* Milk Fever.

PARTURIENT PSYCHOSIS: A clinical term for hyperactivity and increased restlessness in periparturient females. Parturient psychosis may be accompanied by aversive reaction toward offspring (e.g., biting, kicking, pushing, threatening). It occurs most commonly in primiparous females, particularly in swine. Synonym: Puerperal Psychosis.

PARTURITION: The process of giving birth.

PARTURITION PERIOD: A period of time starting with the first symptoms of labor and ending when the afterbirth is eliminated from the mother's body.

PASO (horse): A smooth four-beat gait of Paso horses. The lateral legs are lifted almost simultaneously, but the hindhoof contacts the ground shortly before the forehoof. The sequence of leg movements is similar to that of the slow gait in five-gaited horses, but the action is not as high and the cadence may be faster.

PASSAGE (horse): A dressage gait, performed as a collected and highly rhythmical trot with even beats between diagonally synchronized legs.

PASTERN: The area of the first phalanx.

PASTEURELLOSIS: A disease of cattle, sheep, and swine that may occur in several forms. The most common behavioral symptoms are sudden increase in respiration rate, a rapid rise in body temperature, tremors and shivering, depression, coughing, loss of appetite with symptoms of colic pains, and in pigs, increased tendency to huddle together.

PASTURE: Land with growing plants suitable for grazing.

PATCH AVOIDANCE: A form of selective grazing whereby animals avoid small circular areas of pasture marked by excreta of other grazing animals.

PATCHING: Patch avoidance by grazing animals. (colloquial term)

PATELLAR REFLEX: Extension of the limb in response to a mild blow to the patellar tendon.

PATERNAL: Pertaining to the male parent or sometimes to ancestors on the sire's side.

PATERNAL CARE: Epimeletic behavior of a sire directed toward dependent offspring or young.

PATERNALISM: Control over satisfaction of needs and desires of an individual on its behalf, rather than allowing it to attempt to fulfill such needs and desires itself.

PATERNAL REJECTION: Social rejection of neonatal offspring by their sire.

PATHOGENIC: Referring to organisms that cause a disease.

PATHOGNOMONIC BEHAVIOR: Any behavioral sign symptomatic of a pathological condition of an organism.

PATHOLOGY: The study of the origin, development, and nature of diseases or disorders.

PATIENT, MORAL: *See* Moral Patient.

PATRIARCH: A male that is predominant in the social organization of a given social unit. Such predominance involves alpha social status and arises out of ancestry to other members of the unit.

PATTERN, ACTION: *See* Action Pattern.

PATTERN DISCRIMINATION: Discrimination between complete patterns rather than pattern components.

PAVLOVIAN CONDITIONING: *See* Classical Conditioning.

PAVLOVIAN DISINHIBITION: Recovery of a conditioned response (CR) during extinction training when a novel stimulus occurs simultaneously with the conditioned stimulus. The developing association inhibiting the CR appears to be disrupted in such a situation.

PAWING: A scooping leg movement, generally on the ground. Animals may display pawing in play situations, as warning signals, when searching for food in snow-covered terrain, when in pain, or when frustrated.

PECKING: Making contact with an object or the body of another organism using the tip of the beak. Pecking is performed with a short, quick forward motion of the head and is displayed during investigation following visual inspection, during feeding, during fighting, and during courtship. *Also see* specific pecking—FEATHER.

PECKING RIGHT (poultry): A term occasionally used to refer to unreciprocated pecks directed towards group peers. Pecking right is assumed to be a sign of rigid social dominance or despotism.

PECK ORDER: Social hierarchy in domestic birds.

PEEP: A high-amplitude repeated vocalization produced repetitively by young domestic poultry (and young of a variety of wild avian species) in response to physical discomfort or separation from conspecifics. Each peep is characterized by relatively short duration and primarily descending sound frequency. The peep vocalization also is referred to as a distress call or separation call.

PELT: The skin of a mammal, such as a sheep or mink, with the fur or hair left on.

PELVIC THRUSTING: *See* Thrusting.

PEN: Any housing unit within a barn, separated from others by barriers, which can be occupied by one animal (individual pen) or several animals (group pen). *Also see* specific pen—FLOOR.

PENECTOMY: Surgical removal of a part of a penis. This term also refers to surgical displacement of the penis to prevent natural insemination.

PENIS, BROKEN: *See* Broken Penis.

PERCENTILE: Any unit of distribution based on percentage division. This term is often used in such associations as percentile rank, percentile scale, etc.

PERCEPTION: The process of forming object hypotheses.

PERCEPTION TIME: Time elapsed between the presentation of an object and its perception by the subject.

PERCEPTUAL FIELD: The total array of stimuli, detected through all sensory modes, which an organism is able to perceive.

PERCEPTUAL LEARNING: Learning to perceive and remember or recognize particular features of the environment without apparent reinforcement.

PERCH: Any elevated object upon which a bird stands or sits. Also, the act of standing or sitting on such an object.

PERFORATIVE PERITONITIS: Peritonitis caused by perforation of the wall of the digestive tract by some sharp object that has been ingested. In ruminants, the perforation commonly occurs in the reticular wall of the rumen. Behavioral symptoms depend on the severity and type of perforation but may include cessation of eating, suspended rumination, reduced ambulation, arching of the back, signs of pain during urination and defecation, and sensitivity to palpation of the affected area.

PERIOD: *See* specific period—CRITICAL; REFRACTORY; SENSITIVE.

PERIPARTURIENT BEHAVIOR: Behavioral activities related to and occurring around the time of parturition.

PERIPHERAL NERVOUS SYSTEM: That portion of the nervous system apart from the brain and spinal cord which innervates both viscera and somatic parts of the body.

PERISTALSIS: Directional waves of contraction and relaxation of tubular muscular systems (e.g., peristalsis of the alimentary canal).

PERITONITIS: Inflammation of the peritoneum. Symptoms depend on the severity of the condition, but include reduced feed intake, signs of abdominal pain, constipation, and elevated body temperature. *Also see* specific peritonitis—PERFORATIVE.

PEROSIS (poultry): A disease of chickens, also called slipped tendon, characterized by deformed leg bones and severely hindered motor activity.

PERSONAL DISTANCE: *See* Individual Distance.

PERSONAL SPACE: *See* Individual Distance Zone.

PERSPIRATION: The watery substance secreted by sweat glands in the skin;

the secretion of such watery substances.

PERVERTED APPETITE: A general term occasionally used to refer to pica, infantophagia, cannibalism, or coprophagia. (colloquial term)

PESSAGE (horse): A dressage exercise of the Spanish High Riding School in which the horse raises the anterior portion of its body by supporting itself on its hindlegs which are spread apart and slightly bent at the hocks, and the forelegs are folded underneath the chest. The pessage is performed in the same way as the levade except that the anterior of the horse is raised higher.

PHALLECTOMY: Surgical trimming or amputation of the penis to prevent intromission. Phallectomy is used to develop teaser bulls for estrus detection.

PHARYNGITIS: Inflammation of the pharynx. Behavioral signs are coughing, strained swallowing indicative of pain, and cessation or reduction of eating and rumination.

PHASE, *See* specific phase—APPETITIVE; CONSUMMATORY; POSTRECEPTIVE; PRE-RECEPTIVE; RECEPTIVE; REFRACTORY.

PHENOMENON, WHITTEN: *See* Whitten Effect.

PHENOTYPE: The manifest characteristics of an organism that develop through interaction of the genotype with the environment.

PHEROMONE: A chemical substance secreted and perceived by individuals of the same species. Pheromones are of two types: signaling pheromones, which induce a behavioral response, and priming pheromones, which induce physiological changes. Pheromones are important in sexual behavior.

PHIMOSIS: Narrowing of the lumen of the prepuce that reduces or prevents protrusion of the penis.

PHLEGMATIC: When applied to animals, referring to an apathetic temperament. This term derives from classical categorization of human character types. Antonym: Choleric.

PHOBIA: An excessive, persistent, and apparently unsubstantiated fear specific to given stimuli, situations, or circumstances.

PHONATION: Any vocalization.

PHORETIC BEHAVIOR: The use of one organism by another organism as a means of transport.

PHOTOPERIODICITY: Regular, cyclic alternation between periods of daylight and darkness.

PHOTOPIC VISION: Vision mediated by cones.

PHOTORECEPTOR: A sensory receptor that responds to a certain spectrum of electromagnetic radiation.

PHOTOTAXIS: Taxis related to light.

PHOTOTROPISM: Orienting response to a light source. The response can be toward the light (positive phototropism) or away from the light (negative phototropism).

PHYLOGENIC: Referring to the evolutionary development of a species. *Compare:* Ontogenic.

PHYSIOLOGICAL AGE: An assessment or rating of age based on the condition of physiological systems. *Compare:* Chronological Age.

PHYSIOLOGY: The biochemical processes in living organisms; the scientific study of such processes.

PIAFFE (horse): A dressage gait performed as a highly rhythmical trot on the spot with even beats between diagonally synchronized legs.

PIA MATER: The innermost vascular membrane which covers the brain and spinal cord.

PICA: Abnormal appetite for unusual and often inappropriate feed, e.g., dirt, hair, feces, etc.

PIEBALD (horse): A black coated horse with white spots.

PIECEMEAL ACTIVITY: No longer in general use. The term was used to refer to an organism's responsiveness only to parts of its surroundings or to specific attributes of a complex stimulus. Piecemeal activity can be induced when parts of the environment or attributes of a complex stimulus are substantially altered or missing.

PIG: Young swine, usually below 75 kg body weight. This term is being used increasingly for all categories of swine.

PIGLET: A young pig during its first two months of life.

PIGLET SCOUR: *See* Colibacillosis.

PILING (poultry): Extremely dense accumulation of individuals in a small area so that some may be buried under others. Piling is observed primarily in poultry and is caused by a serious environmental inadequacy, such as inappropriate temperature, strong draft, or limitation of an essential resource, or by events inducing fear and escape responses, such as unusual light or unfamiliar noise, a predator, or an action of a careless attendant. Piling, if not immediately corrected, leads to suffocation of the buried individuals.

PILOERECTION: Erection of hair. Piloerection commonly is a thermoregulatory response and also may be observed during aggressive interactions.

PINNA: *See* Auricle.

PINTO (horse): A horse with spotted, piebald, or mottled coat color.

PIPPED EGG (poultry): An egg that is visibly cracked or perforated by the beak of its hatching chick.

PIPPING: The first cracking of the egg shell by the hatching chick.

PIROUETTE (horse): A dressage maneuver in which the forelegs follow a circle having a radius equal to the distance between the hindlegs and forelegs. Pirouette may be performed in collected walk, collected canter, or passage.

PITCH: The perceived acoustic quality of a given sound wave frequency.

PITUITARY: A gland at the base of the brain consisting of two lobes. The anterior lobe secretes hormones that regulate the function of the adrenal cortex, the gonads, and the thyroid gland. The posterior lobe stores oxytocin and vasopressin secreted by the hypothalamus.

PLACENTA, RETAINED: *See* Retained Placenta.

PLACENTAL STAGE: *See* Labor.

PLACENTOPHAGIA: Eating of own afterbirth by a postparturient female.

PLACER (horse): In dressage, lateral flexion of the head in the direction of movement during performance of the half-pass.

PLANT AVOIDANCE: A form of selective grazing in which animals avoid certain plants of the pasture flora. *Compare:* Defoliation, Progressive Defoliation, Patch Avoidance.

PLANTIGRADE LOCOMOTION: Locomotion conducted on the entire sole of the foot, or foot and hand. *Compare:* Digitigrade Locomotion.

PLATEAU, RESPONSE: *See* Response Plateau.

PLAY BEHAVIOR: A set of activities experienced as pleasurable in themselves by the organism performing them. Play behavior might be directed either toward the surroundings or toward the body of the animal itself. Social play often imitates serious situations (e.g., social conflicts) but without the serious consequences of such situations.

PLEASURE: An emotional state resulting from an enjoyable sensation.

PLEASURE HORSE: A horse exclusively used for recreational purposes, such as racing, driving, or riding.

PLEURITIS: Inflammation of the pleurae or other membranes inside the chest. Behavioral symptoms are perspiration, shivering, breathing difficulties, locomotory hypoactivity, and signs of pain.

PLEXUS: A network of nerves, arteries, veins, or lymphatic vessels related to certain organs or body parts (e.g., cardiac plexus, mammary plexus, sacral plexus, etc.). *Also see* specific plexus—BRACHIAL; LUMBOSACRAL.

PLUMAGE: Feather coverage of a bird's body.

PLURIPAROUS: *See* Multiparous.

PNEUMONIA: An inflammation of the lungs occurring in several forms in all farm animal species. Common behavioral symptoms are accelerating respiration, rising temperature and pulse rate, dilation of nostrils and nasal discharge, loss of appetite, recumbency, suspended rumination (cattle and sheep), shivering (pigs), or extended standing (horses).

PNEUMOPHAGIA: *See* Aerophagia.

POIKILOSTASIS: Maintenance of psychophysiological balance within an organism by locomotion and selection of appropriate environment. *Compare:* Homeostasis.

POIKILOTHERMY: Substantial fluctuation of body temperature coincident with changes in temperature of the environment. *Compare:* Homeothermy.

POISONING: Illness caused by voluntary intake or artificial introduction of toxic chemicals into an organism's body in an amount or at a rate above its level of natural tolerance. The probability of poisoning is highest after a ration or pasture is changed, facilities or fields are sprayed with herbicides, or animals are disinfected or medically treated. Hungry, thirsty, curious, or excited animals are more susceptible to the consumption of poisonous substances.

POLIOENCEPHALOMALACIA: *See* Cerebrocortical Necrosis.

POLLAKIDIPSIA: Unusually frequent drinking without significant change in the total volume of daily intake of liquid.

POLLAKIURIA: Unusually frequent, small-volume urination without significant change in the total daily volume of urine.

POLLARD: A castrate, usually a female chicken whose ovary has been removed; also may refer to a hornless cow or sheep.

POLLING: *See* Dehorning.

POLO (horse): A two-team horseback game played with four players on each team. The objective is to score points by using a long mallet to drive a wooden ball into the opposing team's goal.

POLYANDRY: A mating system in which one female mates with two or more males.

POLYDIPSIA: The consumption of large amounts of liquids (frequently used interchangeably with the term excessive thirst).

POLYESTROUS FEMALES: Females that manifest several estrous cycles per year.

POLYGAMY: A mating system in which a single member of one sex mates with several members of the opposite sex. Polygamy can be either "polyandrous," one female mating with several males, or "polygynous," one male mating with several females.

POLYGASTRIC: Having more than one compartment in the stomach (e.g., ruminants).

POLYGYNY: A mating system in which one male mates with two or more females.

POLYOVULAR: Producing several ova during every ovarian cycle.

POLYPHAGIA: Consumption of an unusually broad variety of foods. *Compare:* Hyperphagia.

POLYPNEA: Abnormally increased respiration rate. *Compare:* Hyperpnea.

POLYSYNAPTIC REFLEX: A reflex in which afferent neurons synapse with intermediate neurons which, in turn, synapse with efferent neurons. *Compare:* Monosynaptic Reflex.

POLYTOCOUS: Referring to animals that typically deliver several offspring at each parturition (e.g., swine, dogs, etc.). Antonym: Monotocous.

POLYURIA: Excessive secretion of urine.

PONS: Part of the hindbrain that forms a bridge of nerve fibers between the two hemispheres of the cerebellum.

PONY: A small horse not taller than 150 cm when mature.

POPULATION: In a broad sense, all contemporary animals of a given species. For practical purposes, the term usually is applied to smaller groups demarcated according to some meaningful criterion (e.g., breed, geographical area, organizational unit, etc.). *Also see* specific population—CLOSED.

POPULATION DENSITY: A quantitative parameter defined as the number of organisms of a given population living within a given territory. *Compare:* Spatial Density.

POPULATION DYNAMICS: The demographic changes in a population over time, or the study of such changes.

PORCINE: Swine or pertaining to swine.

PORCINE STRESS SYNDROME: A syndrome occurring in swine caused by an inherited hypersensitivity to stressful situations. Behavioral symptoms are elevated heart rate, reddening of the skin, muscular rigidity, and in extreme cases, death. Susceptibility can be determined by a halothane test. The meat of animals manifesting porcine stress syndrome is of inferior quality, being pale, soft, and exudative.

PORPHYRIA: Occurrence of an excessive amount of porphyrins in urine, giving it an amber coloration. Affected animals often show retarded growth, and their skin is oversensitive to sunshine.

POSITION: *See* specific position—CLINCH; DOG-SITTING.

POSITIVE FEEDBACK: A process in which the performance of an activity leads to an increase in, or prolongation of, the activity.

POSITIVE REINFORCER: *See* Reinforcer.

POSTERIOR: Situated or directed toward the rear of an organism's body. Antonym: Anterior.

POSTERIOR PRESENTATION: Fetal presentation in which the hind legs and posterior enter the birth canal first.

POST-LEGGED: Pertaining to animals with abnormally straight hind legs.

POST-NATAL: After birth.

POSTPARTUM: The period of time following parturition.

POSTPARTURIENT BEHAVIOR: Behavior of mammalian females displayed after delivery of their offspring. Postparturient behavior characteristically includes puerperal behavior and epimeletic behavior oriented toward offspring (neonatal licking, nursing, protection, play, parental training, and weaning).

POSTRECEPTIVE PHASE: The third and last phase of estrus, characterized by gradual disappearance of sexual behavior. The postreceptive phase begins when the female ceases to be sexually receptive.

POSTURAL REFLEXES: A variety of reflexes by means of an organism maintaining its body in an appropriate position relative to the ground and other elements of its surroundings.

POSTURE: The attitude or position of the body. *Also see* specific posture—AMBIVALENT; HERDING.

POTENTIAL: *See* specific potential—ACTION; GENERATOR; RESTING.

POULT: A young turkey up to the stage of secondary sexual differentiation.

POULTRY: Birds bred for production of eggs or meat.

P

POUNDING (horse): Defective leg action in which the contact of the hooves with the ground appears excessively heavy.

PRACTICE LIMIT: The highest level of skill attainable through practice.

PRECOCIAL: Refers to species whose individuals are sufficiently developed at birth or hatching to see, move in a coordinated fashion, and fend fairly well for themselves. Neonatal individuals of such species require much less parental care than those of altricial species. *Compare:* Altricial.

PRECOCIOUS: More developed than usual for a given stage in the life cycle. Antonym: Altricious.

PRECONDITIONING: Preparation of an animal to cope with changes in environment (social or physical). This preparation may involve exposure to novel feeds, familiarization with different environments, handling, etc. *Also see* specific preconditioning—SENSORY.

PREDATION: A form of interspecies relationship in which the attacker (predator) kills and eats the victim (prey).

PREDATOR: Any animal which preys upon other animals.

PREDATORY: Preying upon other animals.

PREDISPOSITION: An internal condition of an organism that predetermines the efficiency of a given process. Predisposition is used frequently in reference to the genotype (genetic predisposition), conditions fixed at birth (innate predisposition), and propensities influencing ability or efficacy to learn (learning predisposition).

PREENING: An act of integumentary care in birds similar in function to grooming in mammals. Preening is manifested as manipulation of feathers and distribution of secretions from the uropygial gland (preen gland) using the beak, and also as scratching of the body surface with claws or beak.

PREFERENCE: Choice of a certain alternative over another alternative(s). This term generally is used in reference to a specific situation or the outcome of such a situation (e.g., preference test, natural preference, learned preference, spatial preference, social preference, feed preference, color preference, preference index, preference order, etc.). *Also see* specific preference—LEARNED; MATE; NATURAL; OFFSPRING; PARENTAL; SIBLING; SOCIAL.

PREFERENCE INDEX: Any calculated parameter, comparable among observed individuals, reflecting some qualitative or quantitative aspects of a preference response. For example, a color preference index used in genetic selection studies.

PREFERENCE ORDER: Any ranking of alternatives related to a preference situation or resulting from such a situation.

PREFERENCE RATIO: The proportion of occasions in which each alternative of a set of alternatives is chosen (e.g., the proportion of times a particular diet is chosen out of several alternative diets).

PREFERENCE TEST: An experimentally defined choice situation designed to objectively assess preference among alternatives. Synonym: Preference Experiment.

PREGNANCY: The state of a mammalian female characterized by presence of a developing offspring(s) in the uterus. Typical signs of pregnancy are cessation of estrous cycling, increasing docility, tendency to deposit fat, greater susceptibility to fatigue, enlarged abdomen, visible indicators of fetal movements, and in the final stage, parturition. Characteristic periods of progressing pregnancy in cattle and horses are determined by the first absence of expected estrus (negative stage), gradations of the size of the fetus (small sac stage, large sac stage, balloon stage), descended position of the fetus (sinking stage), and parturition (final stage). *Also see* specific pregnancy—DOUBLE.

PREGNANCY DETECTION: Any behavioral observation, chemical test, or physical examination conducted for the purpose of pregnancy diagnosis. The most common procedures used are observation of behavioral changes (cessation of estrous cycling, alteration of maintenance behavior, more careful ambulation), rectal-abdominal palpation, monitoring of hormonal changes, ultrasonic measurements, external tactile examination (bumping), mucus analysis, and observation of changes in body configuration.

PREHENSION: The act of grasping or seizing. Also, the action of the tongue and lips during feed intake of grazing animals.

PREMATURE BIRTH: Birth after an abnormally short gestation. Offspring born prematurely tend to be underweight or underdeveloped in other ways and usually require special care to survive.

PREMATURE EJACULATION: Ejaculation preceding intromission.

PRENATAL BEHAVIOR: Behavior of the embryo or fetus.

PREPARTUM: Before parturition.

PREPARTUM MILK: Milk ejected prior to parturition.

PREPARTURIENT BEHAVIOR: Behavioral actions indicative of impending parturition. Typical preparturient behavior in farm animals comprises reduced ambulatory activity, avoidance of slippery floors, careful locomotion, search for delivery site, nest building in some species (e.g., swine, rabbits), and predelivery excitement followed by increased recumbency with high respiratory rate.

PREPOTENT RESPONSE: The predominant response made when an organism receives stimuli which would elicit two or more different responses.

PREPROGRAMMED LEARNING: Learning that occurs with high probability at a certain age or stage of maturity. Preprogrammed learning (e.g., imprinting, learning of vocalizations) appears to be under strong genetic control and involves information that is predictable and important for fitness.

PREPUCE SUCKING: Sucking on the prepuce of other individuals. It occurs in young group-housed bull calves. Prepuce sucking may lead to infection, abscesses, or stricture of the prepuce causing difficult urination.

PRERECEPTIVE PHASE: The first phase of estrus, characterized by initiation of sexual behavior but without manifestation of sexual receptivity.

PRESENTATION: *See* specific presentation—ANTERIOR; BREECH; CRANIAL; CROUP; DOG-SITTING; DORSAL TRANSVERSE; FETAL; HEAD RESTRAINED; PARTIALLY ANTERIOR; POSTERIOR; UPSIDE-DOWN; UPSIDE-DOWN POSTERIOR; VENTRAL TRANSVERSE.

PRESENTING: Exposure of external genitalia by a female in estrus to a male or sexually cooperative female.

PRESSING: *See* specific pressing—CHIN; HEAD.

PRESSOR: A substance that elevates certain physiological activities such as blood pressure or pulse rate.

PREVENTIVE AGGRESSION: *See* Defensive Aggression.

PRIMARY COLOR: Any color which is important in a given color classification (e.g., color which cannot be broken down into components—blue, yellow, and red). In the Young-Helmholtz theory of color vision, primary colors are those that produce maximal responsiveness in cones of the retina—red, green, and blue.

PRIMARY DRIVE: A drive that occurs without need of prior learning. Synonym: Innate Drive.

PRIMARY PUNISHMENT: A stimulus that without prior experience, is perceived as punishing to an organism, independent of association with any other punishing stimulus. *Compare:* Primary Reinforcer.

PRIMARY REINFORCER: A stimulus that without being previously experienced by the organism has reinforcing properties. A primary reinforcer can be a reward or a punishment.

PRIMARY REWARD: A stimulus that without prior experience, is perceived as rewarding to an organism, independent of association with any other rewarding stimulus. *Compare:* Primary Reinforcer.

PRIMARY SEXUAL CHARACTERISTICS: Distinct sexual characteristics detectable at birth, and essential for reproduction.

PRIMARY SOCIALIZATION: A rapid process that occurs during a species-specific critical period leading to strong social attachment between a neonatal animal and its dam (or other entity) and also, to a varying extent, among siblings. Licking, sniffing, and suckling trigger primary socialization in mammals. *Compare:* Secondary Socialization.

PRIMING: A form of sensitization to a given stimulus through exposure to it prior to its expected occurrence (e.g., exposure of a pregnant gilt to neonatal piglets shortly before she is due to farrow). Priming also may refer to physiological and psychological changes induced by exposure to certain stimuli which prepare the organism to make appropriate responses to other stimuli; e.g., photoperiod change may prime a seasonal breeder to become responsive to sexual stimuli.

PRIMING PHEROMONE: *See* Pheromone.

PRIMIPAROUS: Referring to females having undergone only one cycle of pregnancy and parturition. *Compare:* Multiparous.

PRINCIPLE, NORMATIVE: *See* Normative Principle.

PROACTIVE INHIBITION: *See* Memory Interference.

PROACTIVE INTERFERENCE: *See* Memory Interference.

PROBABILITY: The likelihood or chance of something occurring.

PROBABILITY RESPONSE: The frequency of response in relation to the maximally possible frequency of such a response.

PROBLEM SOLVING LEARNING: Learning in which memory and innovation are integrated by means of cerebration processes to cope with unfamiliar situations or problems.

PROCEDURAL REPRESENTATION: A hypothesis that the operational sequences involved in attaining goals are represented in the mind without propositional conceptualization of objects or relationships in the environment. Procedural representation accentuates the use of available information. *Compare:* Declarative Representation.

PROCEPTIVE BEHAVIOR: Solicitation of a male's attention by a sexually receptive female. *Compare:* Presenting.

PROCESS: A continuous series of successive interdependent events. *Also see* specific process—TRANSFER.

PROCLIVITY: Natural inclination, disposition or strong habitual tendency. Synonym: Propensity.

PRODOME: First symptom(s) of disease.

PRODUCED RESPONSE: In a binary classification relevant to operant conditioning, a response manifested as the performance of a particular motor action. For example if a horse is being trained to jump, the manifestation of the motor action of jumping when the cue to do so is given, is considered a produced response. Antonym: Withheld Response.

PRODUCTION DISEASE: A common designation for health problems caused by imbalance between nutritional intake and production output. Examples of production disease are hypomagnesemia, or hypocalcemia.

PROESTRUS: The first phase of the estrous cycle, characterized by rapid increase in the size of ovarian follicles and the occurrence of early signs of estrous behavior.

PROGENY: Offspring.

PROGENY TESTING: A method used to assess the breeding value of parental individuals by statistical analysis and evaluation of the performance of their progeny.

PROGESTERONE: A hormone produced mainly by the corpus luteum but also found in the adrenal cortex and the placenta of some species. Progesterone interacts with estrogen to regulate sexual behavior of females, prepares the uterus for implantation, and maintains pregnancy.

PROGESTIN: A general designation for any natural or synthetic progesterone-like compound.

PROGRESSIVE DEFOLIATION: A form of selective grazing whereby animals preferentially consume the most palatable parts of some plants before eating less palatable portions. Progressive defoliation is most obvious when pasture flora is inadequate in highly palatable plants.

PROGRESSIVE LAMENESS: Lameness that gradually increases over time, usually caused by inflammation, degeneration of tissue, or neoplasia.

PROGRESSIVE RATIO REINFORCEMENT SCHEDULE: *See* Variable Ratio Reinforcement Schedule.

PROLACTIN: *See* Luteotropin.

PROLAPSE: Protrusion of an organ or part of an organ from its normal position due to inadequate strength of supportive tissue (e.g., anal, cloacal, or vaginal prolapse).

PROLIFIC: Able to reproduce at a rapid rate.

PROLONGED GESTATION: Abnormally extended gestation. Prolonged gestation occurs in some breeds of cattle and sheep and is caused by fetal gigantism (Holstein and Ayrshire cows, Karakul sheep) or deformed fetuses (Jersey and Guernsey cows).

PROMOTANTS, GROWTH: *See* Growth Promotants.

PROPHYLAXIS: Intentional prevention of disease.

PROPRIOCEPTION: Perception of bodily balance, position, and movement.

PROPRIOCEPTIVE NERVOUS SYSTEM: Part of the somatic nervous system that transmits afferent impulses from stimuli occurring inside the body (e.g., in muscles, bones, joints, ligaments). *Compare:* Exteroceptive Nervous System.

PROPRIOCEPTOR: A sensory receptor activated by stimulation arising from muscle, tendon, or ligament action.

PROPULSUS (cattle): A segment of copulatory behavior in males, characterized by a distinct, powerful thrust immediately after intromission, and accompanied by ejaculation.

PROSENCEPHALON: *See* Forebrain.

PROSTAGLANDINS: A group of compounds (all variants of prostanoic acid, a 20-carbon carbonylic fatty acid) produced in many tissues of the body. The major physiological role of prostaglandins involves control of the activity of vascular smooth muscle. Thus, they influence localized blood circulation, uterine contractions, ova and sperm transport, and perhaps, rupture of the follicle. Some other effects of prostaglandins that may not be related to smooth muscle activity have to do with gonadotrophin secretion, ovum maturation, regression of corpora lutea, and release of oxytocin during labor.

PROSTASIS: Any positive or negative effect of social hierarchy ranking on the behavior of an individual organism.

PROVEN SIRE: A sire that meets required criteria based on the progeny test.

PROXIMAL: Near or closer to a point of reference. Antonym: Distal.

PROXIMAL STIMULUS: A stimulus that directly impinges upon sensory receptors. *Compare:* Distal Stimulus.

PROXIMITY EFFECT: The strength of attractiveness or aversiveness of a stimulus contingent on its proximity to an organism, and thus varying with the distance between the organism and the stimulus. In approach-avoidance conflict situations, the approach proximity effect at some point may equal the avoidance proximity effect. In such a case, motivations to approach and avoid are equal, and compromise behavior or displacement behavior often is manifested.

PROXIMOCEPTOR: A receptor that can be activated only by tactile contact.

PRURITUS: Itchiness occurring in most skin diseases caused by parasites.

PSEUDOCONDITIONING: Development of association between an unconditioned stimulus and a previously neutral stimulus that is not paired with the unconditioned stimulus in time but, due to chance, sometimes appears close to it. As a result, a conditioned response becomes elicited upon the occurrence of the previously neutral stimulus.

PSEUDOPREGNANCY: Occurrence of physiological and behavioral symptoms of pregnancy without conception.

PSEUDORABIES: *See* Aujeszky's Disease.

PSS: *See* Porcine Stress Syndrome.

PSYCHE: The soul, spirit, or mind, as distinguished from the physical nature of a body.

PSYCHIC BLINDNESS: Inability of an organism to distinguish the characteristics of a given type of stimulus leading to indiscriminate responsiveness to such stimuli. Psychic blindness is due to disfunction of information processing in the CNS (e.g., serious damage to the temporal lobe) and is not caused by failure of the senses.

PSYCHOGALVANIC RESPONSE: *See* Galvanic Skin Response.

PSYCHOLOGICAL ATTACHMENT: A distinct affection and attraction of an organism toward some object in its surroundings. If the attachment is to a living organism, the phenomenon is also referred to as social attachment.

PSYCHOLOGICAL CASTRATION: Conditioned inhibition of sexual behavior.

PSYCHOLOGICAL ESTRUS: Manifestation of all behavioral signs of estrus, except tolerating mounting.

PSYCHOPHARMACEUTICALS: Chemical compounds that, through their

P

effect on the central nervous system, have the capacity to change behavior.

PSYCHOSIS, PARTURIENT: *See* Parturient Psychosis.

PSYCHOSOMATIC DISEASE: Any physiological disorder having a psychological etiology, e.g., heart failure due to excessive emotional arousal. Ulcer of the esophagus in pigs and abomasal ulcers in cattle are considered psychosomatic diseases.

PSYCHOSOMATIC RESPONSE: Physiological or other bodily responses to psychological phenomena (e.g., increase of oxytocin secretion in lactating female mammals when exposed to vocalization of young).

PUBERTAL BEHAVIOR: Behavioral activities occurring typically during the period of development of secondary sexual characteristics and reproductive maturity.

PUBERTY: The developmental stage of an organism when the gonads start to secrete hormones in amounts that are sufficient to cause maturation of genitalia and appearance of secondary sexual characteristics. In normal circumstances, the period is completed when the organism achieves the capability to produce viable gametes and displays behavior characteristic of sexual maturity.

PUERPERAL BEHAVIOR: Behavior indicative of physical exhaustion immediately following parturition. The characteristic behavioral state is recumbency. The duration of recumbency is influenced by species affiliation and delivery ease.

PUERPERAL PSYCHOSIS: *See* Parturient Psychosis.

PUERPERIUM: The postpartum period extending from the time of expulsion of the placenta until full regeneration of the endometrium, uterine involution and assumption of ovulation and estrous cycling.

PULLET: A female chicken close to sexual maturity.

PULLING, FLEECE: *See* Fleece Pulling.

PULLORUM (chicken): A disease of chickens caused by *Salmonella pullorum*. Behavioral symptoms develop slowly, usually starting with increased and changed vocalization, ruffled feathering, difficulty maintaining positional balance, and yellowish diarrhea.

PUNISHMENT: The application of noxious stimulation. *Also see* specific punishment—PRIMARY; SECONDARY.

PUNISHMENT TRAINING: A type of operant conditioning in which a punishment is directly contingent on the performance of the subject. According to the training objectives, the performance resulting in punishment may be either a produced response or a withheld response.

PUPIL: The circular area in the middle of the iris through which light penetrates to the lens and subsequently is focussed onto the retina. The diameter of the pupil can be controlled by the iris to regulate the amount of light that enters the eye.

PUPILLARY LIGHT REFLEX: *See* Light Reflex.

PUREBRED ANIMAL: An animal whose parents both belong to the same breed.

PYRAMIDAL SYSTEM: The corticospinal tract.

PYREXIA: Abnormally high elevation of body temperature.

QUACKING (ducks): A vocalization of ducks, produced either as a single sound or, more typically, as a short series of low sounds.

QUADRIPLEGIA: Paralysis or paresis of all four limbs. Synonym: Tetraplegia.

QUALITY OF LIFE: State of an organism resulting from total array of conditions affecting its existence. Quality of life may range from negative to positive extremes. Assessment of quality of life requires joint consideration of a broad range of criteria influencing the organism over extended rather than short periods of time. *Compare:* Suffering, Well-being.

QUEEN: A female cat or a fertile female in social hymenoptera (e.g., bees, wasps, ants).

QUICKENING: Visible signs of fetal movement during pregnancy. (colloquial term)

QUIDDING: Abnormal eating behavior, characterized by a repeated sequence of food intake, attempt to chew, and expulsion of food from the mouth. It may be caused by a dental problem or an inability to swallow.

QUIESCENCE: Rest, absence of activity. *Compare:* Refractory Phase.

QUIET ESTRUS: *See* Silent Estrus.

R

RABIES: An infectious disease of the central nervous system, resulting from insertion contact (e.g., by being bitten) with a virus in the saliva of an affected animal. The most common behavioral signs of rabies commence with hypoactivity and seclusion-seeking, proceed to sporadic periods of hyperactivity and aggression, and terminate with paralysis and death.

RACETRACK: A prepared area designated for racing.

RACHITIS: A form of aphosphorosis, caused by vitamin D deficiency or absence of phosphorus in the diet occurring in pigs, foals, calves, and other domestic animals. Behavioral characteristics are progressive lameness, joint swelling, and lateral bending of the limbs.

RACING (horse): *See* specific racing—FLAT; HARNESS.

RACK (horse): An artificially attained fast cadence, four-beat gait of five-gaited horses. Each leg leaves and hits the ground at regular intervals in the sequence, left hind, left fore, right hind, right fore. Synonym: Single Foot.

RADIAL PARALYSIS: Paralysis of the forelimb caused by dysfunction of the radial nerve. It is characterized by lowering of the elbow joint because of loss of shoulder flexion ability. The affected limb is unable to support weight.

RADIOTELEMETRY: Measurement, transmission and recording of information (e.g., behavior records) by means of a remote sensor. A radiotransmitter apparatus transmits radiosignals to a receiver apparatus.

RALES: Chest sounds produced during inspiration and expiration by an animal suffering from respiratory disease.

RAM: A sexually mature male sheep.

RAM BLOCKING: A tendency of some males to place themselves between estrous ewes and other males thus preventing them from mating. Ram blocking is commonly displayed by dominant males and in some situations can have an adverse effect on conception rates.

RAMENER (horse): In dressage, reduction of the angle between the head and neck while the poll is kept at the apex.

RANDOM BREEDING: Sexual reproduction in which gametes originate from animals whose genetic relationship is a random sample of the genetic relationships in the population. *Compare*: Random Mating.

RANDOM MATING: A type of mating system in which individuals of one sex all have equal probability to mate with any given individual of the opposite sex.

149

RANDOM REINFORCEMENT: Reinforcement that occurs unsystematically. Also called aperiodic reinforcement.

RANGE: The distance between the two extreme values of some distribution. *Also see* specific range—ACTIVITY.

RANK, SOCIAL: *See* Social Status.

RANKING: Assignment of elements of a sample or group to positions, classes, etc. in some series or distribution scale.

RANKING SCALE: *See* Ordinal Scale.

RANK ORDER: The arrangement of a sample or group resulting from ranking.

RATE: Quantity of a variable in relation to some other variable or standard (e.g., number of behavioral events per given time period). *Also see* specific rate—HEART; RESPIRATION.

RATIO, SOCIONOMIC: *See* Socionomic Ratio.

RATION, BALANCED: *See* Balanced Ration.

RATIONAL EGOISM: A theory guided by the principle that moral behavior does not require one to sacrifice self-interests, and thus it is morally acceptable to do what a person believes to be in his own self-interest. In relation to the treatment of farm animals some proponents of this theory claim that the promotion of animal well-being is in the best interest of their owner since the highest productivity of farm animals demands highest possible care for these animals. Such a claim is valid only if the highest productivity of an animal is not in a conflict with animal well-being. Should the highest productivity of a farm animal compromise its well-being, such a claim is not valid.

RATIO SCALE: A data scale in which recorded variables are categorized into mutually exclusive groups of equal class interval ranked from an absolute zero point (e.g., speed of locomotion). *Compare*: Interval Scale.

REACTION: *See* specific reaction—ALARM; DEFERRED; FLIGHT.

REACTION TIME: Time between the beginning of stimulation and the initiation of response. *Compare*: Response Latency.

REACTIVE INHIBITION: Diminishing learning rate or lessened response strength with increasing number of preceding responses in a given trial series.

REAFFERENCE: Illusionary movement of objects in the visual field caused by discrepancy between the picture expected by the brain and the image on the retina.

REAFFERENT STIMULATION: A form of proprioception in which mechanical stimulation arises from the organism's own bodily movement.

REARING: Lifting of the anterior part of the body by shifting weight entirely to the hindlegs. Rearing is common in many species during play, aggression, or as a segment of mounting behavior. Rearing of a ridden horse may

R

indicate refusal or be a startle response and, if excessively high, may result in falling over backward, dangerous to both rider and horse.

REAR-LEG LIFT (horse): Partial lifting of a rear leg indicative of readiness to kick. Rear-leg lift is considered to be a threat signal.

RECEPTIVE PHASE: The second phase of estrus, occurring prior to but closely linked with ovulation. The receptive phase is characterized by willingness of an estrous female to engage in mating.

RECEPTIVITY, SEXUAL: *See* Sexual Receptivity

RECEPTOR: A sensory nerve ending that responds to a stimulus and initiates the afferent message to perceptual centers. Receptors commonly are defined according to specified functions, such as photo-, mechano-, chemo-, thermoreceptors, etc. *Also see* specific receptor—CONTACT.

RECUMBENCY: *See* Lie and specific recumbency—LATERAL; SEMILATERAL; VENTRAL.

REDIRECTION: Transfer of behavior from the stimulus that motivates it to an alternative stimulus. Redirection may occur when some factor inhibits the animal's response toward the eliciting stimulus, and the behavior is then directed toward the alternate stimulus.

REDUCED APPETITE: A clinical term for lower than normal intensity of eating or reduced feed intake.

REDUCTIONISM: A theory that postulates that all phenomena (including biological ones) can be fully explained by physical processes.

REFLEX: An intrinsic response mediated by a reflex arc and requiring no or minimal cerebration. *Also see* specific reflex—ANAL; CARDIAC; CORNEAL; COUGH; CROSSED; DEFLATION; EMETIC; ERUCTATION; EXTENSOR; FLEXION; GRASPING; INFLATION; LACRIMAL; LIGHT; MICTURITION; MONOSYNAPTIC; MYOTATIC; NASAL; PATELLAR; POLYSYNAPTIC; RIGHTING; SALIVARY; SCRATCH; SUCKLING; SWALLOWING; TOLERANCE.

REFLEX ARC: A neurological unit consisting minimally of a receptor, an afferent neuron, an efferent neuron, and an effector. More complex reflex arcs involve additional intermediary neurons.

REFLEXES: *See* specific reflexes—ATTITUDINAL; POSTURAL; RESPIRATORY.

REFLEX LATENCY: *See* Reflex Time.

REFLEXOGENOUS ZONES: Bodily zones where mechanical stimulation motivates reflex actions.

REFLEX OVULATION: *See* Induced Ovulation.

REFLEX TIME: Reaction time for a reflex action. *Also see* Reaction Time.

REFRACTORY PERIOD (neuron): The time subsequent to membrane depolarization during which resting potential is restored. An action potential cannot occur during the refractory period.

REFRACTORY PHASE: The final phase of an operant behavioral cycle. The refractory phase follows a consummatory phase and is characterized by voluntary cessation of consummatory responses. *Compare*: Behavioral Cycle.

REFUSAL (horse): Disobedience (e.g., when the horse refuses to jump an obstacle).

REGRESSION: In a biological sense, a return to some original or more primitive form. In a statistical sense, the rate of change of a dependent variable per scale unit of an independent variable.

REGURGITATION: Flow of stomach contents back from the stomach to the mouth. Regurgitation facilitates remastication of boluses of feed in ruminants and provision of food for offspring in canids and some species of birds.

REIN BACK (horse): A dressage maneuver in which the horse moves equilaterally backward. The legs are diagonally synchronized, but the forehooves leave and hit the ground slightly before the hindhooves.

REINFORCED AGGRESSION: Any aggression ascertainably influenced by reinforcement.

REINFORCEMENT: The effect of a reinforcer whereby performance of a particular behavior is induced, eliminated, altered, or maintained at a constant level. Also, the act of applying a reinforcer. *Also see* specific reinforcement—ACCIDENTAL; APERIODIC; INCIDENTAL; RANDOM; SPURIOUS; YOKED.

REINFORCEMENT INHIBITION: Temporary suppression of a conditioned response occurring when reinforcement trials are repeated in close succession over an extended period of time.

REINFORCEMENT SCHEDULE: A predetermined plan according to which a reinforcing stimulus is to be presented under experimental circumstances. *Also see* specific reinforcement schedule—CONTINUOUS; FIXED INTERVAL; FIXED RATIO; INTERMITTENT; MULTIPLE; PROGRESSIVE RATIO; VARIABLE INTERVAL; VARIABLE RATIO.

REINFORCER: A stimulus that occurs or is presented when a particular response is manifested and that influences the probability of the future performance of the response. A "positive reinforcer" increases this probability, while a "negative reinforcer" does the opposite. *Also see* specific reinforcer—CONDITIONED; NEGATIVE; POSITIVE; PRIMARY; SECONDARY; UNCONDITIONED.

REJECTION: *See* specific rejection—MATERNAL; PARENTAL; PATERNAL; SOCIAL.

RELATIONSHIP: *See* specific relationship—INTERSPECIES; SOCIAL.

RELATIVISM, MORAL: *See* Moral Relativism.

RELAXANT: A chemical compound that temporarily reduces tension in specific tissues (e.g., muscle relaxant).

RELAXATION: To enter into or maintain a restful state characterized by lessened work output, reduction of the level of tonus in tissue (e.g., muscular relaxation) or diminished level of mental arousal (mental relaxation).

RELAXIN: A proteinous compound isolated from the corpus luteum and placenta that causes relaxation of pelvic ligaments and dilation of the uterine cervix prior to parturition.

RELEARNING: Reinstitution, through a learning process, of previously learned acts which for some reason have become extinct.

RELEASER: *See* Sign Stimulus and specific releaser—SOCIAL; TIME.

RELIABILITY, OBSERVER: *See* Observer Reliability.

REM-SLEEP: Sleep characterized by rapid eye movement (hence REM) and often irregular respiration and irregular heart beat with detectable activity of the brain resembling a state of arousal. REM-sleep is also called paradoxical sleep or desynchronized sleep. *Compare*: SW-Sleep.

RENVERS (horse): An inverse travers.

REPERTOIRE, BEHAVIORAL: *See* Behavioral Repertoire.

REPORT, BRAMBELL: *See* Brambell Report.

REPRESENTATION: Subjective mental simulation of objects, events, or procedures mediated by the application of perceptual processes. Two alternative forms of representation have been proposed, declarative representation and procedural representation. *Also see* specific representation—DECLARATIVE; PROCEDURAL.

REPRESENTATION LATENCY: The time taken by an animal to solve a spatial problem posed to it. Representation latency has been used to assess the complexity of the perceptual processes involved in representation.

REPRODUCTION: The process of copying something. In biology, commonly referring to production of a new generation by parental organisms.

REPRODUCTIVE BEHAVIOR: Behavioral actions involved in the sexual reproduction of an organism, including search for a mate(s), courtship, copulatory behavior, and parental care.

REPULSION: The phenomenon of being stimulated to move away from or avoid some object, organism, or action because of its inherent characteristics. Antonym: Attraction.

REPULSION SIGNAL: Any vocal, visual, olfactory, or other sign, or combination of such signs, broadcast by an organism to repel other organisms from its spatial proximity or to discourage them from attaining such proximity.

RESIDUAL MILK: Milk that remains in the mammary gland after nursing or milking. Also called complementary milk.

RESILIENCE, BEHAVIORAL: *See* Behavioral Resilience.

RESPIRATION: Inspiration and expiration of air.

RESPIRATION CYCLE: The recurrent sequence of the two behavioral segments involved in breathing, i.e., inspiration and expiration.

RESPIRATION RATE: The number of respiration cycles per given unit of time. The typical range of cycles per minute is 8-10 for horses, 10-30 for cattle, 15-25 for sheep and pigs, and 20-30 for goats.

RESPIRATORY REFLEXES: Reflexes that comprise the respiratory activity of an animal. Included are the inflation reflex, deflation reflex, and cough reflex.

RESPONDENT: A behavioral action(s) interpreted in the context of its immediate cause.

RESPONDENT CONDITIONING: *See* Classical Conditioning.

RESPONSE: Any observable muscular, glandular, or neural activity that results from stimulation. *Also see* specific response—ALARM; BEHAVIORAL; COOLING; CONDITIONED; CONSUMMATORY; FOLLOWING; IMPLICIT; ORIENTING; OVERT; PRODUCED; SELECTIVE; SENSORY ADAPTATION; SERIAL; SPONTANEOUS; UNCONDITIONED; WITHHELD.

RESPONSE LATENCY: The time between the presentation of a stimulus and the beginning of a response. Synonym: Lag, Latency Time, Latent Time. *Compare*: Reaction Time.

RESPONSE MECHANISM: A hypothetical unit of the psychophysiological system involved in the manifestation of a particular response.

RESPONSE PATTERN: An organized sequence of responses having a specific role. Response pattern is essentially a subcategory of behavioral pattern.

RESPONSE PLATEAU: A leveling off in the rate of a process or flattening of a response curve (e.g., plateau in a learning process).

RESPONSE PROBABILITY: The proportion of responses occurring relative to the total number of response opportunities (a measure of response strength).

RESPONSE STRENGTH: Any measure of magnitude or intensity of a response.

RESPONSIVENESS: The capacity or willingness of an organism to respond to a given stimulus. It may be measured by some assessment of response latency, response strength, etc.

REST: Interruption of training or work to avoid or recover from fatigue, exhaustion or overtraining.

RESTING: A behavioral state characterized by cessation or reduction of movement and lowered expenditure of bodily energy to avoid or recover from exhaustion. It is often accompanied by lowered level of alertness.

RESTING AREA: A location used regularly by one or more animals when resting. Characteristically, resting areas provide safety, protection against adverse weather conditions, or a good vantage point for surveillance of the surroundings. The term also is used to refer to animal holding facilities or pens designed to be used by animals when resting.

RESTING POTENTIAL: The electric potential across the membrane of an inactive neuron caused by differences in ion concentration at equilibrium (i.e., the membrane is polarized between inside and outside). Resting potential enables the generation of an action potential when depolarization occurs.

RESTLESSNESS: A behavioral state characterized by sustained arousal and motor activity. Typical actions include short and repeated periods of pacing, scratching, ground scratching, head tossing or jerking, grooming or preening, higher occurrence of urination, frequent alteration between standing and lying, and increased ambulation.

RESTRAINT: Any technique used to temporarily discourage or prevent unwanted movement. Restraint is used for examination, surgery, convalescence, breeding, and safe handling of animals.

RESTRICTED FEEDING: *See* Limited Feeding.

RETAINED PLACENTA: A placenta that remains within the uterus for an extended length of time (usually over 24 hours) after expulsion of the fetus. Retained placenta is relatively frequent in cattle and its occurrence increases with multiple birth, dystocia, abortion, prolonged gestation, and uterine prolapse.

RETENTION: Persistence of information in the memory of an organism.

RETICULAR ACTIVATING SYSTEM: A structure of nerve fibers located in the thalamus that regulates transmission of sensory information to various centers of the cerebral cortex, and controls consciousness.

RETINA: The inner layer of the posterior part of the eyeball, able to receive images of the visual field projected through the crystalline eye lens. The visual receptors in the retina are rods and cones.

RETRACTED EARS (horse): Ears flattened along the dorsal portion of the head. Retraction of ears is a common display during agonistic encounters.

RETROACTIVE INHIBITION: *See* Memory Interference.

RETROACTIVE INTERFERENCE: *See* Memory Interference.

RETURNED APPETITE: A clinical term for resumption of normal intensity of eating or normal feed intake after a period of inappetence or reduced appetite.

REVERSAL TRAINING: A series of discrimination trials in which the correct solution is regularly alternated between opposite possibilities. The learning set to be acquired involves the principle of regular alternation between objects to be chosen as the correct solution.

REWARD: A stimulus or situation that occurs contingent upon performance of a particular behavior and that has pleasant or satisfying characteristics for a given organism. *Also see* specific reward—ADVENTITIOUS; PRIMARY; SECONDARY.

REWARDED AGGRESSION: *See* Offensive Aggression.

REWARD LEARNING: *See* Reward Training.

REWARD TRAINING: A type of operant conditioning in which a reward (positive reinforcer) is directly contingent on the performance of the subject. According to the training objectives, the behavior resulting in reward may be either a produced response or a withheld response.

RHEOTAXIS: Taxis related to direction of current in a fluid.

RHINITIS: An acute or chronic inflammation of nasal passages. Behavioral symptoms are snorting and head shaking in horses and cattle, sneezing and rubbing of the snout in pigs, and watery nasal discharge, followed by thick nasal discharge. Synonym: Catarrh, Coryza, Ozena.

RHOMBENCEPHALON: *See* Hindbrain.

RHYTHM: *See* specific rhythm—BIORHYTHM; EXOGENOUS; ENDOGENOUS; FREE-RUNNING.

RICKETS: *See* Rachitis.

RICOCHETAL LOCOMOTION: Locomotion conducted by jumps or hops.

RIG: A cryptorchid horse. (colloquial term)

RIGHTING REFLEX: A postural reflex to achieve an upright body position. The term also refers to intrauterine positional adjustment of the fetus preceding parturition.

RIGHT, PECKING: *See* Pecking Right.

RIGHTS: Entitlements of an individual that ought to be respected by moral agents. These entitlements originally were considered to apply exclusively to humans, but some animal welfare philosophers have argued that they also apply to animals according to the same rationale by which they pertain to humans.

RING WOMB (sheep): Resistance or inability of the cervix to dilate, thus obstructing natural delivery of the fetus. (colloquial term)

RITUALIZATION: An evolutionary phenomenon whereby behavioral actions having direct biological functions in themselves become modified, often into stereotypic form, and incorporated into standardized sequences (modal action patterns) that have a communicative function frequently quite different from the original function of their respective components.

RITUALIZED AGGRESSION: A term used occasionally to refer to assertion of social dominance without fighting.

RITUALIZED BEHAVIOR: Behavior, often eclectic in nature, performed in a symbolic, stereotyped manner which is part of behavior patterns typical of, and often specific to, a species (e.g., courtship feeding in some birds).

ROADSTER (horse): A horse trained, exhibited, or competing in the pulling of bikes or road wagons. Roadsters are judged working at a walking pace and at slow, medium, and full speed trot.

ROAR (horse): A low to medium amplitude sound of variable duration (0.2-1.5 sec) emitted during breathing with the mouth slightly open. It is caused by reduced control of the vocal cords and may be a consequence of damage to

R

the laryngeal nerve. Roaring may develop after severe pneumonia and can be corrected surgically.

ROD: A photosensitive cell in the retina. Rods function as the only receptors of light in dim illumination. *Compare:* Cone.

RODEO (horse): Formalized exhibition of cowboy skills, including saddle bronc riding, bareback riding, bull riding, calf roping, steer wrestling, team roping, and cutting horse contest.

ROD VISION: *See* Scotopic Vision.

ROLLING (cattle): *See* Tongue Rolling.

ROLLING (horse): An activity characterized by lying down and rolling onto the back accompanied by rubbing movement against the ground. Rolling is considered to be comfort related. The term also refers to side-to-side movement of the body during running in wide-fronted and wide-based horses.

ROLLING (poultry): *See* Egg Rolling.

ROOST: Any elevated object used as a resting place by birds. In animal agriculture, any object constructed by humans for such use by birds. Also, the act of resting on such objects.

ROOSTER: A mature male chicken.

ROOTING (swine): Digging the ground with the snout.

ROPE-WALKING (horse): Defective leg action in which the moving leg travels in a semicircle around the supporting leg and the hooves are planted too close to the horse's midline.

ROTATION: A twisting or turning movement around the longitudinal axis of a body or limb.

ROUGHAGE: Any food for livestock such as hay, silage, or straw that is high in cellulose fiber. Roughage has a relatively low energy and protein content.

ROUTE HORSE: A horse trained for long distance racing. (colloquial term)

r-SELECTION: Natural selection that favors low parental investment in individual offspring and production of large numbers of offspring. *Compare*: K-Selection.

RUB: To press some portion of the body against a blunt object, or another portion of the body, and generate friction by moving one or the other while they are in sustained contact. Rubbing commonly is a component of integumentary care. Excessive rubbing may also occur in association with skin disease. In some circumstances, rubbing may become stereotyped and thereafter may lead to damage of the integument (e.g., tail-rubbing of horses).

RUBBER, BODY: *See* Body Rubber.

RUBBING: *See* Chin Pressing and specific rubbing—HOOF; TAIL.

RUFFLING (chicken): A postural display characterized by ruffled feathers, extended neck, and a short period of intense shaking of the whole body. Ruffling is considered to be a comfort movement.

RUMINAL TYMPANY: *See* Bloat.

RUMINANT: An animal having a stomach system with four compartments (rumen, reticulum, omasum, abomasum) adapted for the digestion of food containing high levels of cellulose. Ruminant animals, such as cattle, sheep, and goats, characteristically spend a substantial portion of the day engaged in rumination.

RUMINATION: Regurgitation, mastication, and reswallowing of boluses of feed by ruminant animals. *Also see* specific rumination—SHAM.

RUNNING: Relatively fast locomotion of an organism on the ground in which propulsive force derives from the action of legs.

RUNNING HORSE: A thoroughbred horse.

RUNNING WALK (horse): An artificially attained fast four-beat gait. The legs leave the ground at different times, and the cadence of all four beats should be even in the sequence, left hind, left fore, right hind, right fore (the running walk differs from the rack in lower action of all legs).

RUNT: An individual that is small and underdeveloped in comparison with litter mates or peers of equal age.

S

SADDLE GROOMING: Allogrooming of the saddle region conducted mutually by two animals.

SALIVARY REFLEX: Expression of saliva from the salivary glands into the mouth in response to the presence or expectation of food.

SALMONELLOSIS: A bacterial disease of cattle, sheep, pigs, horses, poultry, and several other species. Behavioral symptoms are occurrence of scours and bloody discharges from the intestines, and progressive weakness followed in some cases by death. Pigs show ear discoloration, tendency to rest in close social proximity, and difficulty maintaining balance when ambulating.

SAND-BATHING: *See* Dust-bathing.

SANGUINE: Having an alert and confident temperament. The term is derived from classical categorization of human character types.

SARCOCYSTOSIS: A disease of cattle, sheep, pigs, and horses caused by organisms of the genus *Sarcocystis*. Behavioral symptoms may include anorexia, lameness, and hypersalivation accompanied by anemia, ulceration, hemorrhage, fever, and abortion.

SARCOPTIC MANGE: A contagious disease of the integument caused by itch mites, *Sarcoptes scabiei*. Symptoms include pruritis and eczema. Synonym: Scabies.

SATIETY: A state of complete satisfaction of a given desire.

SATISFIER: Any stimulus or experience that meets a need or desire.

SATURATION: An attribute of color, referring to the intensity of hue.

SAVAGING: Destructive behavior that causes serious injury or death of other organisms. This term is commonly used in reference to biting of neonatal piglets by their dam. *Compare*: Overlying.

SCABIES: *See* Sarcoptic Mange.

SCALE: *See* specific scale—ABSOLUTE; BEHAVIOR; CLASSIFICATORY; INTERVAL; NOMINAL; ORDINAL; RANKING; RATIO.

SCALPING (horse): Defective leg action in which the hoof of the hindleg hits the coronet of the foreleg.

SCENT MARKING: Olfactory markings that delineate territory, designate trails, or attract sexual partners.

SCHEDULE-INDUCED BEHAVIOR: Activity of an organism that is causally linked to a reinforcement schedule imposed on the organism. Schedule-induced behavior may be manifested during or between the intervals in which reinforcement is applied.

SCHEDULE, REINFORCEMENT: *See* Reinforcement Schedule.

SCHOOLING: The formation and maintenance of socially coordinated groups in fishes. Occasionally this term also refers to training assistance provided by parents or guardians (e.g., obedience training).

SCIENCE: *See* specific science—ANIMAL; ANIMAL WELFARE.

SCIENTIFIC EVIDENCE: Evidence based on methodically collected and objectively analyzed observations.

SCORE, OBJECTIVE: *See* Objective Score.

SCOTOMA: A small area in the retina of the eye that is partially or completely blind.

SCOTOPIC VISION: Vision in dim light; dark-adapted vision mediated by rods.

SCOUR, PIGLET: *See* Colibacillosis.

SCOUR, WHITE; *See* Colibacillosis.

SCOURING (horses): Outbursts of undesirable activities during training session or work. (colloquial term)

SCOURS: Prolonged diarrhea in animals.

SCOUTING BEE: *See* Exploratory Bee.

SCRAPIE (sheep): A clinical term for a contagious, fatal, ovine spongiform encephalopathy. Symptoms include pruritus and muscular incoordination.

SCRATCHING: Any repeated or rhythmical rubbing action against objects in the surroundings (e.g., wall, ground surface, etc.) or rubbing action between two parts of an animal's own body (e.g., foot against neck). *Also see* specific scratching—HEAD.

SCRATCH REFLEX: A reflexive response to itching irritation of the integument. This reflex occurs in two stages involving specific location of the irritating stimulus, and application of abrasive force to it by means of some oscillating body movement.

SCROTUM SUCKING: Sucking of the scrotum of other individuals. Scrotum sucking may occur in young, group-housed animals and may cause abrasion and infection.

SEARCH: Goal-oriented investigation.

SEARCH IMAGE: Temporary selective attention to relevant exogenous stimuli developed through reinforcement associated with searching behavior. Search image commonly is used in the context of foraging behavior and is postulated to aid the animal to discern what it searches for from the background complex of environmental stimuli.

SEASON: Estrus. (colloquial term)

SEASONAL BREEDING: Breeding that occurs exclusively and regularly during a certain part of the year. Some breeds of sheep, as well as cats and dogs, are seasonal breeders.

SECONDARY DRIVE: A drive whose occurrence is dependent on some prior learning experience. Synonym: Learned Drive.

SECONDARY PUNISHMENT: A stimulus that acquires punishing properties through association with some primary punishment. *Compare*: Secondary Reinforcer.

SECONDARY REINFORCER: A stimulus that acquires reinforcing properties (either positive or negative) through association with some primary reinforcer.

SECONDARY REWARD: A stimulus that acquires rewarding properties through association with some primary reward. *Compare*: Secondary Reinforcer.

SECONDARY SEXUAL CHARACTERISTICS: Anatomical, physiological, and behavioral characteristics that differ between the sexes and develop most rapidly during the pubertal period but are not directly related to reproductive ability.

SECONDARY SOCIALIZATION: Socialization that occurs after the critical period for primary socialization. Secondary socialization takes longer to establish, is not associated with any discernable critical period, and leads to relationships that are less filial in nature.

SECOND-ORDER CONDITIONING: Conditioning in which an animal learns to manifest a conditioned response (CR) to a conditioned stimulus (CS) through the reinforcing properties of an intermediary conditioned stimulus. For example, if an animal salivates (CR) in response to the sound of a bell (CS) because it has learned an association between the ringing of a bell and food (UCS), the animal then may learn to salivate (CR) when a light is turned on if the light stimulus is presented in proper temporal conjunction to the sound of a bell, in the absence of food.

SECRETION: Production and release of substances by glands.

SEDATION: Reduced excitability. Also, the process of reducing excitability (e.g., by treatment with a sedative).

SEDATIVE: A chemical compound that temporarily reduces nervous excitability.

SEE: To perceive stimuli through the visual sense mode.

SEGMENT, BEHAVIORAL: *See* Behavioral Segment.

SEGREGATION: *See* Social Segregation.

SEIZURE: *See* Convulsion.

SELECTION: In a reproductive context, any process which leads to differential reproductive success among individuals or groups. If such a process is under direct human control, it is referred to as artificial selection; otherwise, it is called natural selection. *Also see* specific selection—ARTIFICIAL; BIDIRECTIONAL; DIRECTIONAL; DISRUPTIVE; DIVERGENT; FOOD; INTERSEXUAL; INTRASEXUAL; K; KIN; NATURAL; r; STABILIZING; SEXUAL.

SELECTIVE ATTENTION: A tendency to be more attentive to a particular stimulus (or type of stimulus) than to other stimuli present at the same time.

SELECTIVE RESPONSE: A response chosen by an organism from a number of possible alternatives.

SELF-AWARENESS: A state of being cognizant of one's own existence, and thus aware of one's own action and place in the environment.

SELF-FEEDING SYSTEM: A feeding system that permits ad libitum intake by individual animals whenever feed is available.

SELF-FELATOR: Oral self-stimulation of the penis. It occurs frequently in the Canidae and Felidae families.

SELF-GROOMING: *See* Grooming.

SELF-LICKING: *See* Licking.

SEMANTIC COLOR: A color which has specific social meaning.

SEMICIRCULAR CANALS: Canals situated in the inner ear which apparently serve to facilitate balance.

SEMILATERAL RECUMBENCY: Lying during which the posterior part of the body is in lateral contact with the ground while the thorax contacts the ground only in the sternal area with forelegs folded to each side or one or both extended cranially. One or both hindlegs in semilateral recumbency may be folded or extended. Semilateral recumbency is very common in four-legged farm species. If rumination is conducted while lying, it is done so in lateral recumbency.

SENESCENCE: Physiological changes in an organism as it reaches old age.

SENILITY: Mental and physical impairment due to senescence.

SENSATION: Any experience resulting from stimulation of sensory receptors, sensory nerves or sensory areas in the brain. In a biophysical context, sensation often refers to a basic element of excitation theoretically independent of learning, motivation, or social circumstances.

SENSE DATUM: The immediate effect of the stimulus experienced by the sensory receptor.

SENSE MODALITY: *See* Sense Mode.

SENSE MODE: A specialized system of receptors, nerve fibers and perceptual centers that allows an organism to obtain information about its environment. Sense modes are generally identified as auditory, gustatory, olfactory, proprioceptive, tactile, and thermal.

SENSE ORGAN: A body structure that facilitates reception of stimuli by sensory neurons (e.g., ear, eye).

SENSES: The total array of sense modes of an organism.

SENSITIVE PERIOD: *See* Critical Period.

SENSITIVITY: The capacity to respond to a given stimulus. For example, an organism with normal hearing is sensitive to sound waves; a deaf organism is not. *Also see* specific sensitivity—ABSOLUTE; LIMINAL; TERMINAL.

SENSITIZATION: The process of becoming more responsive to a given stimulus with practice or number of trials.

SENSORIMOTOR ARC: *See* Neural Circuit.

SENSORY ADAPTATION: The change in sensitivity of a sense mode to stimulation. Decreased sensitivity usually is called negative adaptation, and increased sensitivity, positive adaptation.

SENSORY ADAPTATION RESPONSE: *See* Sensory Adaptation.

SENSORY DEPRIVATION: An absence or insufficiency of exogenous stimulation of an organism which reduces its perceptual opportunities and causes aberrations in its behavioral inventory.

SENSORY DISCRIMINATION: The process or ability to recognize differences between stimuli.

SENSORY NEURON: An afferent neuron that has the capacity to conduct impulses from sensory receptors.

SENSORY PRECONDITIONING: A phenomenon in which an originally neutral stimulus becomes able to elicit a conditioned response (CR) even though no direct contingency is established between the stimulus and an unconditioned stimulus (UCS). By pairing the presentation of two neutral stimuli for a number of trials, and then pairing one of the neutral stimuli with a UCS so that it becomes a conditioned stimulus, the other stimulus may also become a conditioned stimulus able to elicit the CR.

SENTIENCE: Capacity for sensing or feeling. In the context of behavior, sentience may refer to capability of self-awareness and emotion.

SENTINEL BEHAVIOR: Sensory focussing of a guarding animal(s) beyond the periphery of its group, apparently to detect predators or territorial incursions by neighboring groups.

SEPARATION: *See* Social Separation.

SEPARATION CALL: A call produced when an individual is removed from its group or becomes separated from conspecifics or other social partners. Such a vocalization is thought to be an attempt to re-establish communication and, ultimately, contact with peers.

SEQUENCE, BEHAVIORAL: *See* Behavioral Sequence.

SEQUESTRATION CALLS: Warning calls broadcasted to keep others from approaching too closely to an individual distance zone, guarded young, or guarded territory.

SERIAL RESPONSE: A response pattern in which the responses occur in a fixed, temporal sequence.

SERPENTINE (horse): A dressage maneuver in which the horse starts at the middle of one short side of the arena and travels along a path having wide oscillations from one longitudinal side to the other, ending at the middle of the opposite short side of the arena.

SERVICE: Insemination.

SERVING: Mating in cattle and goats.

SETTER, TIME: *See* Time Setter.

SETTING HEN: A broody hen incubating eggs.

SEX: Binomial classification of organisms (i.e., female or male) according to characteristic morphology and behavior, chromosomal array, or gamete type produced.

SEX CHROMOSOME: *See* Chromosome.

SEX DETERMINATION: The consequence of the coming together of a pair of sex chromosomes when two gametes unite to form a zygote. In mammals, females are homogametic, having two X-chromosomes, and males are heterogametic, having one X-chromosome and one Y-chromosome. In birds males are homogametic (ZZ chromosomes), and females are heterogametic (ZW chromosomes). The combination of sex chromosomes leads to development of the sexual characteristics and gender-specific behavioral patterns.

SEX-INFLUENCED: Refers to characteristics having a greater tendency to occur in one sex than in the other because the genes underlying such characteristics are dominant in one sex and recessive in the other.

SEX-LIMITED: Refers to characteristics affecting only individuals of one sex.

SEX-LINKED: Referring to characteristics controlled by genes located on the sex chromosomes.

SEXUAL AGGRESSION: Aggression displayed toward a conspecific of the same sex during competition for a sexual partner, or aggression displayed toward a partner during sexual interaction.

SEXUAL BEHAVIOR: Activities related to sexual sensation, sexual arousal, and sexual gratification.

SEXUAL BIMATURISM: A phenomenon in which males and females of a given species or breed differ in the age of sexual maturity.

SEXUAL CHARACTERISTICS: Any anatomical, physiological, or behavioral characteristics distinctly differentiating the sexes. *Compare*: Primary Sexual Characteristics and Secondary Sexual Characteristics. *Also see* specific sexual characteristics—PRIMARY; SECONDARY.

SEXUAL CONGRESS: *See* Congress.

SEXUAL CROUCH (poultry): Cooperative posture of a female to facilitate mounting by a sexual partner. Sexual crouch is performed as bending and partial spreading of legs, lowering of the body, mild relaxation of the wings, and exposure of the cloaca.

SEXUAL LICKING: *See* Licking.

SEXUAL MATURITY: An ontogenic stage during which the organism produces viable gametes.

SEXUAL RECEPTIVITY: Willingness of a female to engage in mating interaction with a sexual partner. The period of sexual receptivity in farm animals occurs typically during estrus. Reliable behavioral signs of sexual receptivity are display of cooperative postures and standing while mounted.

SEXUAL SELECTION: Selection of sexual partners mediated by social interactions within sexes, e.g., aggression to gain access to females (intrasexual selection), or between sexes, e.g., choice among courting males by a female (intersexual selection).

SEXUAL SENSATION: Sensation resulting from the stimulation of erogenous zones or organs.

SHAKING: An apparently voluntarily induced short period of rapid spasms of the muscles that control the integument. Shaking is commonly performed after dust-bathing, rolling on the ground, or when the coat becomes wet.

SHAKING, HEAD: *See* Head Shaking.

SHAKING DANCE (bees): Rapid dorso-ventral shaking of the abdomen by worker bees, frequently in the proximity of the queen.

SHAM CHEWING: Chewing actions performed without the presence of food in the oral cavity. If performed excessively, sham chewing may be accompanied by hypersalivation. Sham chewing occurs most frequently in confined pigs fed a highly concentrated diet and is considered to be a vacuum activity.

SHAM DUST BATHING: Dust bathing activity performed without the presence of litter material or other manipulable floor substrate (e.g., on plastic or wire floors). Sham dust bathing is considered to be a vacuum activity.

SHAM RUMINATION: Display of mandibular movements typical of rumination without presence of feed in the mouth. Sham rumination is a vacuum activity manifested by ruminants whose diet is lacking in roughage.

SHEARING: Artificial removal of fleece from the body of a sheep or any other wool-producing animal.

SHEEPDOG: A dog bred or trained to assist a shepherd in control of a flock of sheep.

SHELTER-SEEKING BEHAVIOR: Any action indicating a tendency to seek out environmental conditions that provide protection against danger or discomfort.

SHIN HITTING (horse): Defective leg action in which the hoof of the foreleg hits the canon or shin of the hindleg.

SHIVERING: Rapid clonic spasms that prevent or reduce hypothermia by generating body heat. Shivering also can be a reaction to extreme fear or excitement.

SHOCK: The act of causing or the state of experiencing a sudden and extreme physical, physiological, or psychological disturbance resulting in partial or complete incapacitation.

SHOEING: The attachment of a protective material to the bottom of the hoof of an animal to prevent its wear and injury.

SHORT HORSE: A horse trained to race over short distances; a sprinter. (colloquial term)

SHORT-TERM MEMORY: Memory that is characterized by relative limitation in the amount of information that can be retained and by short retention time. Short-term memory is thought to have a different control system than long-term memory.

SHOULDER: The area around the scapula bone.

SHOULDER-IN (horse): A dressage maneuver in which the horse moves forward, with its body slightly bent around the right leg of the rider. The horse's head should point to the right of the direction of movement.

SHOVEL BEAK (chicken): A tongue deformity that causes feeding difficulty. The occurrence of this deformity is higher in chickens fed on an all-mash diet. (colloquial term)

SHRIEK (chicken): A high-amplitude vocalization, akin to the peep, produced by a chick experiencing intense discomfort or fear.

SHY BREEDER: Mature animals of either sex that have very low reproductive success.

SHYING (horse): Spontaneous and often unpredictable backward and sideways movements indicative of startlement. A symptom of an excitable and inexperienced horse.

SHYNESS, WATER: *See* Water Shyness.

SIBLING PREFERENCE: Preference of an individual to associate with siblings. Such a preference may be expressed at several levels, e.g., preference for a sibling versus a nonsibling conspecific, for a sibling of the same age versus one that is a different age or for certain individuals versus others in a litter.

SIBLINGS: Organisms that have one or both parents in common.

SIB TESTING: A method used to assess breeding value of individuals by statistical analysis and evaluation of the performance of their siblings.

SICKLE DANCE (bees): Movement in a semicircle, with the bee turning and retracing its path a number of times. The opening of the semicircle points to the source of food.

SICKLE-HOCKED: Referring to animals with crooked hocks.

SIDE STICK: A simple restraint device consisting of a stick connected at one end to a halter and at the other end to a girth tied around the chest. A side stick restricts horizontal movement but not vertical movement of the neck, and thus interferes little with eating and drinking.

SIFTING (ducks, geese): A form of feed intake by water fowl in shallow water, characterized by straining water through the beak.

SIGHT: Vision. Also, the characteristics of a particular visual image, or to obtain such an image.

SIGHT AVERSION: A phenomenon similar to taste aversion except that the relevant stimuli by which the food to be avoided is identified are visual in nature. Birds such as chickens and pigeons can learn aversions quickly to the sight of food. *Compare*: Taste Aversion.

SIGNAL: Any visual, tactile, auditory, chemical, or other sign emitted to convey information. Signals may be transmitted between organisms, or between systems within organisms. *See* Expression and specific signal— ALARM; APPEASEMENT; ATTRACTION; BROADCAST; COMPOSITE; DIRECTED; DISCRETE; DISTRESS; DOMINANCE; GRADED; GREETING; REPULSION; SOCIAL STATUS; STATUS; SUBORDINATION; THREAT; WARNING.

SIGNALING PHEROMONE: *See* Pheromone.

SIGN STIMULUS: The stimulus in an array of simultaneous stimuli that consistently evokes a specific response. This term usually is applied to stimuli that activate instinctive behavior.

SILENT ESTRUS: Estrus not accompanied by behavioral symptoms of estrus.

SILENT HEAT: *See* Silent Estrus.

SILENT OVULATION: The first ovulation in the life of a female. This ovulation usually is unaccompanied by detectable behavioral symptoms of estrus. *Compare*: Silent Estrus.

SIMULATION: The act of feigning.

SINGING (chicken): Low frequency sounds of uneven duration produced by mature hens. Singing is assumed to be associated with contentment and satiation.

SINGLE FOOT: *See* Rack.

SINISTRAL: Pertaining to the left side of the body. Antonym: Dextral.

SIRE: The male parent.

SITTING: Body position in which the posterior of the body trunk is in contact with the ground and supports most of the body weight. *Also see* specific sitting—MOTIONLESS.

SKELETAL MUSCLE: *See* Striate Muscle.

SKILL: Aptitude in the performance of behavior.

SKIMMING: A short period of running on the water and vigorous flapping of the wings by some species of waterfowl prior to take-off from water surfaces.

SKINNER BOX: An apparatus designed by B.F. Skinner, used to study operant conditioning. The apparatus generally is automated and allows for presentation of predetermined stimuli and reinforcements. The response operandum usually is a lever or a disk.

SKIN PINCH: Grasping of a fold of skin of an animal (generally on the neck) to control its attention and thus facilitate handling ease.

SLEEP: A state of bodily rest manifested by inhibition of voluntary activities and partial suspension of consciousness. *Also see* specific sleep—REM; SW.

SLING: A device consisting of a wide canvas or leather girth and placed below an animal's trunk. When connected to some overhead support, a sling can safely hold the animal above the ground or support it in a normal standing position.

SLIP (poultry): An incompletely castrated male. (colloquial term)

SLIPPER CLAW: An overgrown claw causing walking difficulty, particularly on slippery surfaces. (colloquial term)

SLIPPING, HALTER: *See* Halter Slipping.

SLOW GAIT (horse): An artificially attained four-beat gait performed by five - gaited horses. Both legs on one side of the body are lifted simultaneously but, due to the high action of the foreleg, the hindhoof hits the ground slightly earlier, resulting in two distinct intervals between beats. Synonym: Stepping Pace.

SMOOTH MUSCLE: Muscle tissue of the intestines, bladder, blood vessels, etc. Smooth muscle also is called organic, nonstriated, or involuntary muscle. *Compare*: Striate Muscle.

SNAP: A sudden, quick, and usually noisy biting action.

SNARE: A simple device, used mostly for brief restraint of pigs, consisting of a loop made of strong rope or smooth cable which can be retracted through a pipe and thus tightened around a pig's snout.

SNATCH: A sudden, quick grasping and holding of objects with the mouth.

SNEEZING: A sudden, powerful expulsion of air through the nose and mouth induced as a reflex spasm. *Compare*: Nasal Reflex.

SNIFFING: A series of brief inhalations of air occurring during olfactory investigation.

SNORT (horse): A medium to high amplitude, rapidly pulsed sound of short duration (0.5-0.75 sec) produced with the mouth closed by powerfully blowing air through the nostrils, which vibrate strongly. It often is emitted during olfactory investigation, immediately after a startling experience or during a work-out under saddle or when pulling, characteristically at the beginning of a new task. Snorts generally are more frequent in young horses.

SNORT (sheep): A medium amplitude sound of very short duration (0.2-0.5 sec) produced with a closed mouth. Snorts often are emitted by startled animals or, if accompanied by stamping with one foreleg, are a threat signal.

SOCIABILITY: Tendency to seek and maintain the company of peers. This term sometimes is used to refer to an animal's attachment to humans.

SOCIAL ACCEPTANCE: A positive or at least neutral attitude toward other individuals or groups.

SOCIAL ADAPTATION: A process of conforming to the behavioral requirements or habits of a given social environment.

SOCIAL AFFILIATION: *See* Affiliation.

SOCIAL ATTACHMENT: A distinct affection and/or attraction of an organism toward some other organism(s) of its own or different species. *Compare*: Psychological Attachment.

SOCIAL BEHAVIOR: Activities directed toward and influenced by other members of a social unit.

SOCIAL BOND: *See* Bond.

SOCIAL DEPENDENCE: Reliance on other organisms for assistance without which survival would be difficult or impossible.

SOCIAL DISTANCE: The distance that two or more individuals or groups maintain between themselves. This distance generally will fall within a certain range determined by the combined effects of cohesive and dispersive social forces characteristic of the species involved, as well as by the given environmental circumstances.

SOCIAL DOMINANCE: Ascendency of an individual over another individual.

SOCIAL DRIFT: An unexplained or random change in the social behavior of a group.

SOCIAL DYNAMICS: The development and changes of social relationships and the interplay of social behavior among organisms during a given period of time.

SOCIAL FACILITATION: A phenomenon in which the behavior of an animal reflectively increases the occurrence of the same behavior among its social partners.

SOCIAL FAMILIARIZATION: Acquisition of information about conspecifics, or members of other animal species, facilitating their subsequent correct identification. The process of social familiarization among adult domestic animals starts characteristically with distal investigation (visual, auditory, olfactory) followed by an approach response and proximal investigation (tactile, gustatory). Among neonatal animals, social familiarization begins with proximal investigation. *Compare*: Socialization.

SOCIAL HIERARCHY: The rank order of individuals in some social unit according to their dominance/subordinance relationships. Synonym: Social Rank Order.

SOCIAL INSTINCT: A natural tendency to form and live in distinct social units.

SOCIAL INTERACTION: Any behavioral interchange between two or more organisms.

SOCIAL ISOLATION: Removal of an individual from its social unit and housing it to prevent any physical contact with conspecifics. Social isolation is a common strategy to minimize spread of infections. *Compare*: Social Separation, Social Segregation.

SOCIALITY: The process and properties of social existence.

SOCIALIZATION: A process of mutual familiarization between organisms which, if successful, leads to full social integration and relatively stable social structure. *Also see* specific socialization—PRIMARY; SECONDARY.

SOCIAL LICKING: *See* Licking.

SOCIAL ORGANIZATION: The total network of social relationships among members of a given social unit. Social organization may be relatively stable if a social hierarchy is well established, or labile if the social hierarchy is being formed or readjusted due to changing composition of the unit.

SOCIAL PREFERENCE: Preference related to social interactions or the result of such interactions.

SOCIAL RANK: *See* Social Status.

SOCIAL RANK ORDER: *See* Social Hierarchy.

SOCIAL REJECTION: Exclusion of an individual from a social unit by persistent manifestation of aversion toward it or refusal to provide care.

SOCIAL RELATIONSHIP: Any association between two individuals in which social behavior is manifested.

SOCIAL RELEASER: Any sign stimulus produced by an individual or group that elicits manifestation of social behavior patterns by conspecifics.

SOCIAL ROLE: A pattern of behavior that an individual is reinforced to adopt as a member of a group.

SOCIAL SEGREGATION: Subdivision of a larger social unit into two or more smaller units according to specific criterion (e.g., age, gender, reproduction stage, production performance). *Compare*: Social Isolation, Social Separation.

SOCIAL SEPARATION: Voluntary or imposed severance of physical contacts between two or more individuals. *Compare*: Social Isolation, Social Segregation.

SOCIAL STATUS: The rank in the social hierarchy attained by an individual as a consequence of its interactions with other members of its social group. Synonym: Social Hierarchy Status.

SOCIAL STATUS SIGNAL: Any behavioral display or sign indicative of the social position of the performer. Status signals are displayed essentially as dominance signals or subordination signals and facilitate maintenance of social order within a group.

SOCIAL STATUS TRANSFER: Transfer of the social status of an individual from one group to its new group. Such a transfer could be classified as positive if the animal attains a similar position in the new group, or negative if the position in the new group is inversely correlated with the previous one.

SOCIAL STRUCTURE: *See* Social Organization.

SOCIAL SUBORDINANCE: Acceptance of the ascendency of another individual(s).

SOCIAL TOLERANCE: The ability to accept the proximity of other organisms when using some common resource.

SOCIAL UNIT: Any set of animals in which each individual has a social relationship with all other individuals.

SOCIETY: A group of individuals organized into some socially cooperative unit.

SOCIOBIOLOGY: The systematic study of social behavior from the viewpoint of genetic fitness.

SOCIOGRAM: An ethogram of social behavior indicating preferences and aversions as well as dominance-subordinance relationships among members of a group.

SOCIOMETRY: The study of the dynamics of social relationships among individuals within a group or among groups within a given area, using statistical, mathematical, or topographical methods.

SOCIONOMIC RATIO: The proportional relationship between different categories of animals within a given social unit, such as female to male ratio in breeding herds or flocks.

SOCIOTOMY: Splitting of a colony. Swarming in bees is a natural form of sociotomy.

SOLITARY: A term referring to an individual which characteristically lives alone and infrequently associates with other conspecifics. The same term may be applied in a general sense to a species whose individuals are typically solitary. Antonym: Gregarious.

SOMATIC NERVOUS SYSTEM: A portion of the peripheral nervous system that innervates skeletal muscles, bones, joints, ligaments, skin, ears, and eyes.

SOMATOTROPIN: A hormone of the anterior lobe of the pituitary gland that stimulates growth of the body. High level of secretion in young animals results in excessive growth; low level of secretion results in reduced growth. Somatotropin promotes synthesis and retention of protein in the body. Synonym: Growth Hormone.

SOMESTHETIC NERVOUS SYSTEM: That portion of the peripheral nervous system involved in sensory reception and bodily sensation.

SOMNOLENCE: Sleepiness.

SOMNOLENT MALE: A sexually hypoactive male. When exposed to an estrous cow, a somnolent bull may lay his chin on the hindquarters of the cow and stay behind her with closed eyes without any attempt to mount.

SONG: Vocalization consisting of a species-specific sequence of notes. This term most commonly refers to vocalizations of animals having some distinct social context (e.g., courtship song).

SONOGRAM: Record of a sound registering tone amplitude, frequency, and intensity.

SONOGRAPH: An instrument that analyzes sounds.

SOPOR: A deep or profound sleep.

SOPORIFIC: Causing or tending to cause sleep.

SORREL (horse): A yellowish-brown or yellowish-red shade of chestnut coat color.

SOUL: A vaguely defined nonphysical entity which, according to different schools of thought, is present exclusively in humans, in living objects or in all objects in the universe. In ancient times some changes or motions were observed to result only from changes in other things, but in living objects some changes could not be traced to changes in other things. To explain such changes various theories were devised and that which allegedly was responsible for such changes was called "psyche" or "soul." In Aristotle's view, all living things possess a soul. Plants have the most rudimentary souls which accounts for the plant's capacity to obtain nourishment and grow. Animals have more complex souls which enable the animal not only to obtain nutrition but also to sense or perceive objects and to move. Humans have the most complex soul as only they have the power of reason. *Also see* specific soul—ANIMAL.

SOUND: Sensation mediated by the auditory sense mode due to vibratory stimulation caused by pressure waves (sound waves) transmitted through air or some other medium. *Also see* specific sound—STENOSIS.

SOW: A female swine after delivery of her first litter of offspring.

SPACING BEHAVIOR: Behavioral activities by which organisms establish and maintain appropriate distances among group members or between adjoining groups. *Compare*: Social Distance, Individual Distance Zone, Group Distance Zone.

SPAN: A dyad of animals harnessed together.

SPARKING-OVER: A term arising out of the concept of action-specific energy.

It was suggested that when the energy of one drive cannot be discharged because behavior specific to that drive is thwarted, the energy build-up eventually reaches a threshold that allows transfer into another drive. Such energy transfer was called sparking-over and was said to give rise to allochthonous behavior.

SPASM: Convulsive contraction of muscles.

SPATIAL DENSITY: The number of organisms per given unit of space. In farm animals, spatial density is commonly expressed as the number of animals per square meter of pen floor, per hectare of field, or per square kilometer of territory. Spatial density also is called stocking density, stocking rate, crowding density, grazing density, or population density.

SPAY: To surgically remove ovaries.

SPECIATION: The evolutionary process of species differentiation resulting from isolation of the gene pools of two or more populations mediated by integrated adaptive changes in behavior (e.g., mate, food, or habitat selection), physiology, and morphology.

SPECIES: A category of the taxonomic scale that is essentially the basic unit of taxonomic classification. A species generally is considered to consist of all organisms capable of inter-breeding without human intervention and producing viable, fertile offspring.

SPECIESISM: A human attitude of bias toward the interests of the human species relative to the interests of other species. The term was coined by Peter Singer and Richard Ryder.

SPECIES-SPECIFIC BEHAVIOR: Behavior that is specific to a particular species of animal. Vocalizations, threat signals, and reproductive behavior patterns frequently are species-specific.

SPECIFIC HUNGER: The development of a preference for food containing specific nutrients in which an animal is deficient. Specific hunger is thought to characterize a special case of operant conditioning in which reinforcement involves alteration in the animal's physiological state and occurs usually a long time after the operant response has been performed. *Compare*: Nutritional Wisdom.

SPECIFIC PATHOGEN FREE: Referring to a prophylactic herd formation technique in which fetuses are surgically removed from the uterus shortly before the time of natural parturition to prevent their contamination by certain pathogens when passing through the birth canal. Thereafter, the herd is maintained as a closed herd with, generally, natural parturition.

SPEEDY CUTTING (horse): Defective leg action in which the hoof of the foreleg hits the pastern or fetlock of the hindleg.

SPINAL ACCESSORY NERVE: The eleventh cranial nerve which innervates the muscles of the neck and shoulders.

SPINAL CORD: The caudal continuation of the medulla oblongata segmented by extensions or pairs of spinal nerves. The spinal cord receives afferent nerve fibers through dorsal roots and extends efferent nerve fibers through ventral roots.

SPINAL NERVES: Nerves that emerge from the spinal cord and mediate both afferent and efferent connection with specific areas of the body.

SPINAL REFLEX: Any reflex principally mediated through the spinal cord.

SPLAY-FOOTED: Characterized by the feet pointing to the side (outwards). This walking defect is also called toe-out or slew-foot.

SPLAY LEG (swine): Myofibrial hypoplasia of neonatal piglets, characterized by paralysis of the hind legs or, occasionally, all four legs. (colloquial term)

SPLENIC APOPLEXY: *See* Anthrax.

SPLIT ESTRUS: A term used to refer to extended period of estrus interrupted by short intervals of anestrus. Split estrus occurs most commonly in mares.

SPONGIFORM ENCEPHALOPATHY, BOVINE: *See* Bovine Spongiform Encephalopathy.

SPONTANEOUS ACTIVITY: *See* Spontaneous Response.

SPONTANEOUS BEHAVIOR: Behavior induced by endogenous stimulation without detectable external causation.

SPONTANEOUS OVULATION: Ovulation that is independent of copulation or artificial stimulation of the vagina and cervix (e.g., cows, mares, sheep, and goats exhibit spontaneous ovulation). *Compare*: Induced Ovulation.

SPONTANEOUS RECOVERY: Reappearance or increase in strength of an extinguished conditioned response after a period of rest from extinction training.

SPONTANEOUS RESPONSE: A response initiated without observable stimulation.

SPREADER INJURY: Injury caused when the forelegs or hindlegs splay laterally beyond their normal range of motion. Spreader injuries are more likely to occur when footing is slippery, particularly if animals are caused to move too quickly or to make sudden movements and changes in direction.

SPRINGER: A female in an advanced stage of pregnancy. (colloquial term)

SPUR (poultry): A horn-like protuberance occurring in males and some females on the inner side of the shank and occasionally used as a weapon.

SPURIOUS REINFORCEMENT: *See* Incidental Reinforcement.

SQUATTING STANCE: *See* Mating Stance.

SQUAWKING (chicken): A variety of calls made by adult birds in response to predators or when trapped or handled in such a way as to cause fear. Squawking differs between sexes and breeds, but usually is produced as a relatively high frequency sound of variable duration.

SQUEAL (horse): A high amplitude sound of intermediate duration (0.5-1.5 sec). This sound frequently is emitted during agonistic interactions and is produced with the mouth open and retracted at the corners, ears laid back, and head raised or turned sideways toward the opponent, and often immediately precedes striking out with the legs.

SQUEAL (swine): An extended sound (0.5-2.0 sec) of both high amplitude and high frequency produced with an open mouth, indicative of a high level of excitement, fear, or pain. Squeals are more frequent in young animals.

SQUEEZE: A device used to mechanically immobilize conscious animals for a short time to allow some treatment to be performed on them, e.g., dehorning, hoof trimming, branding.

STABILIZING SELECTION: Selection such that organisms at or near the mean of the population frequency distribution for the selected trait are favored relative to more extreme individuals. Stabilizing selection tends to reduce additive variance and lead to a narrow range of phenotypes around the mean.

STAG: A castrated mature male exhibiting secondary sexual characteristics.

STALE: An animal that has not been used for work for a lengthy period of time so that its physical condition or ability to obey commands have become substandard.

STALL: A place in a barn designed to be occupied by a single animal or, exceptionally, a dam with her neonatal offspring.

STALL, FREE: *See* Free Stall.

STALL, TIE: *See* Tie Stall.

STALLION: A full-grown sexually mature male horse.

STALL KICKING (horse): Repeated kicking with hindlegs or forelegs against walls, doors, or other parts of a stall. Stall kicking generates a disturbing noise in the barn and is considered to be a dangerous vice as it often leads to leg injuries. Synonym: Stall Knocking.

STALL KNOCKING (horse): *See* Stall Kicking.

STALL WALKING: Pacing or stereotyped circling in a stall. Primarily referring to such behavior in horses. Stall walking is considered to be a nuisance vice.

STAMPING: Striking of the sole of a forefoot on the ground. Stamping is performed by sheep and goats and is a threat signal.

STANCE, MATING: *See* Mating Stance.

STAND: To assume or maintain an upright position on extended legs. The behavior patterns involved in standing up show a high degree of uniformity within species, but may be modified by environmental circumstances (e.g., restriction of movement, injury). If an animal has been lying for a long time, it may manifest leg stretching and ventroflexion, frequently followed by defecation, upon standing.

STANDARDIZATION GROUP: A sample of subjects from which representative scores of behavioral tests are obtained and against which specific scores of other subjects can be compared.

STANDARD STIMULUS: A given stimulus against which other stimuli are compared. Also, a given value of a stimulus against which other values are compared.

STANDING ESTRUS: *See* True Estrus.

STANDING HEAT: *See* True Estrus.

STAPES: *See* Tympanic Ossicles.

STARTER DIET: A diet specifically formulated for young animals. This term is most commonly used in reference to poultry and swine.

STARVATION: Food deprivation.

STASIS: A static interval in behavior dynamics, or a resting state.

STATE, BEHAVIORAL: *See* Behavioral State.

STATUS SIGNAL: *See* Social Status Signal.

STATUS, SOCIAL: *See* Social Status.

STEAMING-UP: Increased feeding of concentrates to dams prior to delivery in order to build up reserves for lactation. (colloquial term)

STEEPLECHASE (horse): A race over artificial or natural obstacles.

STEM, BRAIN: *See* Brain Stem.

STENOSIS SOUND: Any of a variety of sounds produced during breathing as a result of partial obstruction of air flow in the upper respiratory tract. Stenosis sounds often are indicators of inflammatory constriction of respiratory pathways. The sounds may consist of hissing, purring, roaring, snoring, snuffling, or whistling.

STEPPING PACE (horse): *See* Slow Gait.

STEREOSCOPIC VISION: Binocular vision. Slightly different images of the visual field on each retina facilitate perception of depth.

STEREOTAXIS: A locomotion in which contact with some solid body is the main directive factor.

STEREOTYPED BEHAVIOR: Behavior repeated in a very constant way. The term generally is used to refer to behavior that develops as a consequence of a problem situation such as extended social isolation, low level of environmental complexity, deprivation, etc. Stereotypy also may arise from genetic predispositions, or from disease of, or damage to, the brain.

STEREOTYPED MOVEMENT: Any motor activity classified as being an integral part of stereotyped behavior.

STEREOTYPED RESPONSE: Any stereotyped behavior causally linked to some activating stimulus.

STEREOTYPY: *See* Stereotyped Behavior.

STERILITY: Natural or artificially induced permanent inability to produce viable gametes. *Compare*: Infertility.

STEROID: A general designation for any of a large group of hormones that contain a hydrogenated cyclopentenophenanthrene-ring system, e.g., corticosteroids, androgens, estrogens, progestins.

STIMULATION: Generation of neural activity by a stimulus. In the context of experimental behavior, the term stimulation also applies to an attempt to generate such neural activity. *Also see* specific stimulation—ENDOGENOUS; EXAFFERENT; EXOGENOUS; REAFFERENT; VESTIBULAR.

STIMULUS: Any property of the environment (external or internal) that induces neural activity. *Also see* specific stimulus—ADEQUATE; COMPOUND; CONDITIONED; DISTAL; GENERALIZED; IMPERCEPTIBLE; INADEQUATE; INCIDENTAL; INDIFFERENT; KEY; NEUTRAL; ORIGINAL; PROXIMAL; SIGN; STRUCTURED; SUPERNORMAL; TERMINAL; TEST; UNCONDITIONED.

STIMULUS BARRIER: The hypothetical psychophysiological mechanism by which an organism reduces or prevents overstimulation. *Compare*: Stimulus Filtering, Sensory Adaptation.

STIMULUS FILTERING: Processing of afferent stimulation such that some or many sense data do not sufficiently influence an animal to evoke a response. Stimulus filtering enables an animal to attend to a particular stimulus out of the array of stimuli impinging on it at any given time.

STIMULUS RELEVANCE: A phenomenon of learning in which individuals demonstrate an innate psychological propensity to associate certain effects with certain stimuli (relevant stimuli). Other stimuli and effects may not be associated nearly so easily. For instance, in many species, sickness may readily be associated with the taste of food eaten previously (taste aversion), but taste of food may not be as easily associated with some other form of negative reinforcement, such as electrical shock.

STIMULUS-RESPONSE THEORY: A theory stating that in classical conditioning, a conditioned response (CR) develops because it is associated with a reward. The CR, therefore, is not equivalent to an unconditioned response to an unconditioned stimulus. *Compare*: Stimulus Substitution Theory.

STIMULUS SUBSTITUTION THEORY: A theory stating that in classical conditioning, the formation of an association between a conditioned stimulus (CS) and an unconditioned stimulus (UCS) involves the psychological substitution of the CS for the UCS. The conditioned response, therefore, is manifested as if it was an unconditioned response, without regard for the consequences or reinforcement of the action. *Compare*: Stimulus-Response Theory.

STIRRUP: *See* Tympanic Ossicles.

STOCHASTIC: Referring to mathematical probabilities that take into account the possibility that chance may affect relationships between variables. Antonym: Deterministic.

STOCK: A general term for livestock. Also to place animals in facilities set aside for their use.

STOCK, GENETIC: *See* Genetic Stock.

STOCKING DENSITY: *See* Spatial Density.

STOCKING RATE: *See* Spatial Density.

STOICAL: Phlegmatic, emotionless, or apparently indifferent to pain or pleasure.

STOTTING: A type of locomotion, occasionally displayed by sheep and some wild species of ungulates, in which movement forward is conducted in a series of long jumps with all four legs leaving the ground at the same time (or almost at the same time).

STRAIN: A subgroup of a breed comprising animals having genetic and phenotypic characteristics in common that are distinctive in some way from the other members of the breed.

STRAINBREEDING: Sexual reproduction in which gametes originate from animals belonging to the same strain. Strainbreeding accumulates genes characteristic of a given strain and, if practiced over several generations, will increase the average genetic relationship between individual animals within the strain.

STRAINCROSSING: Sexual reproduction in which gametes originate from animals belonging to two separate genetic strains.

STRESS: Any disruption of an animal's homeostatic equilibrium requiring the animal to make some response to maintain its psychophysiological integrity. *Also see* General Adaptation Syndrome.

STRESSOR: Any stress-inducing agent (e.g., physical injury, fear provoking stimulus, extreme environmental temperature, poor air quality, etc.).

STRESS SYMPTOM: Any sign or behavioral display indicative of the presence of a stressor. The most common symptoms are increased excitability, reduced appetite, displacement activities, stereotypic movements, reduced coordination of locomotion, lethargy, and in an extreme case, death of the organism. Symptoms of stress vary according to the intensity and duration of the stressor and the species, age, experience, and health of the stressed organism.

STRETCHING: A muscular activity, characterized by brief, forceful extension of limbs or other parts of the body. Stretching is considered to be a comfort movement.

STRIATE MUSCLE: Skeletal or heart muscle, characterized by microscopic dark and light latitudinal stripes. Striate muscle also is called striped muscle or, somewhat incorrectly, voluntary muscle. *Compare*: Smooth Muscle.

STRIDE: The distance between midpoints of two consecutive prints of the same foot.

STRIDOR: A high-pitched sound produced during respiration and indicative of deformed laryngeal configuration or partial obstruction of the respiratory pathways.

STRINGHALT (horse): Sudden, excessive contraction of the flexor muscles of the hock causing the hindleg to rise higher than normal for a given gait. It occurs more frequently in older horses (over 5 years of age) and usually starts to develop as a light jerk of the limb when the animal is turning.

STRIPED MUSCLE: *See* Striate Muscle.

STRUCTURE, SOCIAL: *See* Social Organization.

STRUCTURED STIMULUS: A compound stimulus whose components are controlled by the experimenter.

STRUTTING (chicken): A threat display characterized by a chase with raised ruff and tail feathers, and trailed wings.

STUPOR: A state of complete or substantial diminution of sensitivity.

STUPOROUS DEPRESSION: A severe form of depression in which most behavioral activities are inhibited.

SUBCONSCIOUS: A term applied to internal processes analogous to conscious processes but of which the organism itself is not consciously aware.

SUBDOMINANT ACTIVITY: A behavioral action of which the occurrence is determined by motivational competition with an alternative action. The subdominant activity is manifested only when disinhibited due to decline in motivation to perform the dominant activity. *Compare*: Dominant Activity.

SUBJECT: An individual organism subjugated to observations or undergoing experimental or other treatments.

SUBJECT-OF-A-LIFE: A term introduced (T. Regan) to designate an organism that, according to Regan's moral theory, has fundamental moral rights. An organism is a subject-of-a-life if it is aware of its surroundings, possesses a set of psychological characteristics (e.g., memories, emotions, feelings), and has the capacity to initiate actions to realize its preferences.

SUBLIMINAL: Below a given threshold.

SUBLIMINAL PERCEPTION: Subconscious perception of a stimulus that has potential to influence behavior.

SUBLIMINAL STIMULUS: A stimulus so faint or presented so briefly that the organism is unable to consciously detect it.

SUBMISSION: Manifestation of deference to another organism. Submission is not necessarily dependent on the social statuses of the interacting organisms.

SUBORDINANCE: *See* Social Subordinance.

SUBORDINATION SIGNAL: Any behavioral display or sign performed by an organism that is indicative of acceptance of subordinance to another organism. Common subordination signals are spatial avoidance, avoidance of visual or tactile contact, allogrooming of the dominant individual, exposure of vulnerable areas of the body, frozen posture, readiness to accept punishment, or escape from the scene of a threatening encounter or fight.

SUBSTITUTE ACTIVITY: *See* Displacement.

SUCKING: Grasping an object in the mouth and creating a partial vacuum in the oral cavity by coordinated actions of the lips, tongue, and buccal muscles. *Also see* specific sucking—BELLY; EAR; NAVEL; PREPUCE; SCROTUM.

SUCKLING: Acquiring milk by sucking on the teats of a lactating female or nursing device. The term generally refers to the acquisition of milk by young from their dam. As a noun, suckling also refers to a preweaned mammal. *Also see* specific suckling—ADULT.

SUCKLING REFLEX: The sequentially synchronized movements of the tongue and buccal muscles that create a sucking action.

SUDDEN DEATH SYNDROME: Death of an apparently healthy organism.

SUDDEN LAMENESS: Lameness that occurs instantly as a consequence of injury to a limb or limb-related tissues (e.g., injured tendon, ligament, or muscle, dislocated joint, bone fissure or fracture, claw lesion, etc.).

SUDORIFIC: Causing the secretion of sweat.

SUFFERING: A psychological state of a sentient organism resulting from perception of harm. Suffering may arise from physiological disruption, injury, pain, physical handicap, privation, etc., and can only be inferred from observable signs exhibited by an animal. *Also see* specific suffering—UNNECESSARY.

SUN BATHING: Basking in the sun to utilize solar radiation for the purpose of thermoregulation and comfort.

SUN STROKE: *See* Heat Stroke.

SUPERFECUNDATION: Production of a greater than normal number of offspring, generally by means of superovulation and embryo transfer.

SUPERFETATION: The presence of two or more fetuses in the uterus that originated from different ovulation cycles and different conception times.

SUPERNORMAL STIMULUS: A stimulus that has one or more attributes exaggerated above the natural level, thus influencing an organism's response. For example, mild increase in temperature of an artificial vagina tends to reduce the latency of ejaculation during semen collection.

SUPEROVULATION: Production of more ova than normal for a given species during one ovulation cycle. Superovulation can be induced by administration of equine chorionic gonadotropin or follicle stimulating hormone.

SUPPLEMENTAL LIGHTING: Any augmentation in level or duration of ambient illumination (e.g., artificial lighting of a barn interior).

SUPPORTIVE LAMENESS: Lameness that occurs when transfer of weight generates pain in the supporting limb. As a result, an ambulating animal will lift this limb sooner than normal and increase the length of the supportive phase of the opposing limb. It also may place more weight on the opposing limb.

SUPPRESSION: The elimination of a response for a period of time.

SUPRALIMINAL: Above a given threshold.

SURROGATE: A substitute. The term commonly is used in reference to females used as recipients of embryos or to females that nurture and provide care for young in the stead of their dam.

SURVIVAL VALUE: Assessment of a trait (behavioral, physiological, morphological, etc.) with regard to its contribution to the survival of an organism. *Compare*: Adaptive Value.

SUSPENDED ANIMATION: Low vitality in neonatal animals necessitating artificial induction of regular respiration.

SUSPENSORY LAMENESS: Lameness that occurs when lifting and extending a limb generates pain. As a result, an ambulating animal will minimize the lifting and extension action of the affected limb. Suspensory lameness usually is caused by injury or disease in the upper part of the leg.

SWALLOWING REFLEX: A complex set of actions that controls swallowing of food and liquids and prevents food from entering the respiratory tract.

SWARMING (bees): Colony fission in bees. (colloquial term)

SWAYBACK (sheep): A disease of neonatal lambs caused by cerebral demyelination, behaviorally manifested by paralysis beginning in the hind legs.

SWEAT: *See* Perspiration.

SWIMMING: Locomotion of an organism through water or liquids.

SWINE FEVER: *See* Hog Cholera.

SWINE INFLUENZA: An acute, contagious disease caused by myxovirus. Behavioral symptoms are loss of or reduced appetite, breathing difficulties, raised body temperature, and violent coughing when pigs are disturbed or forced to move rapidly.

SWISHING, TAIL: *See* Tail Swishing.

SW-SLEEP: Slow wave sleep characterized by reduced respiratory rate, decreased blood pressure, and delta brain waves. Slow wave sleep is also called deep sleep, delta wave sleep, or nonREM sleep. *Compare*: REM-Sleep.

SYMBIOSIS: A form of interspecies relationship in which organisms of different species mutually benefit each other.

SYMBOLIC ACTION: Any action that does not necessarily serve its original purpose, but has some specific function in the life of an organism (e.g., food presenting or food calling as a part of ritualized courtship displays).

SYMMETRIC GAIT: A gait in which the action of each leg is synchronized with another leg. *Also see* Trot; Pace.

SYMPATHETIC NERVOUS SYSTEM: Part of the autonomic nervous system, branching mainly from the thoracic and lumbar portions of the spinal cord. Sympathetic stimulation increases heart rate and blood pressure, dilates the pupils and bronchi and decreases activity of the digestive tract. *Compare*: Parasympathetic Nervous System.

SYMPATRY: Occupation of the same geographical area by two or more populations. Interbreeding is possible among sympatric populations of the same species.

SYMPTOM: A sign that may be used as subjective evidence about the internal condition of an organism. *Also see* specific symptom—PAIN; STRESS.

SYNAPSE: A junction between two neurons that is able to transmit impulses.

SYNCHRONIZATION: Repeated occurrence of two or more events at the same time. This term commonly refers to the joint appearance of previously independently occurring events. *Also see* specific synchronization—ESTRUS.

SYNCHRONIZER, TIME: *See* Time Setter.

SYNDROME: A set of symptoms that occur together and usually indicate reliably the occurrence of a specific physical or psychological disorder in an organism. *Also see* specific syndrome—BARKER; BULLER-RIDER; CEREBELLAR; CEREBRAL; DUMMY; GASPING TURKEY; KLÜVER-BUCY; LEG WEAKNESS; PORCINE STRESS; SUDDEN DEATH; VIRILE.

SYSTEM: *See* specific system—BEHAVIORAL; CENTRAL NERVOUS; CEREBROSPINAL; EXTRAPYRAMIDAL; NERVOUS; PYRAMIDAL; RETICULAR ACTIVATING.

T

TACHIOSCOPE: An instrument used to study perception, designed for very brief presentation of visual materials.

TACHY-: Prefix relating to swift or rapid actions.

TACHYCARDIA: Excessive heart rate.

TACHYPHAGIA: Excessive speed of eating.

TACHYPNEA: Excessive breathing rate.

TACHYPRAGIA: Excessive rapidity of movements.

TACTILE: Pertaining to the sense of touch.

TAIL BITING (swine): Aberrant behavior consisting of playful mouthing, chewing, or rigorous biting of the tails of other pigs. Tail biting very often leads to serious injury and death of victimized animals.

TAIL CROSS-TIE: A restraint device consisting of a rope that is wrapped twice around the base of an animal's tail. If both ends of the rope are pulled laterally in opposite directions, sideways movements of the posterior part of an animal's body can be inhibited.

TAIL FLAPPING: *See* Tail Wagging

TAIL RESTRAINT (cattle): Mild twisting of the base of the tail by hand to control an animal's attention and promote handling ease.

TAIL RUBBING: Stereotyped rubbing of the tail against walls and other objects in a stall. Primarily used in reference to such behavior in horses. Tail rubbing often is an indication of parasites or skin disease, but also can be a sign of boredom. Tail rubbing is considered to be a nuisance vice.

TAIL SWISHING: Swinging of the tail so that the long hairs at the end of the tail brush against the body of the animal. Tail swishing is performed by horses and cattle and serves as a defensive action against flies.

TAIL WAGGING: Rapid, swinging of the tail from side to side. Tail wagging is performed in a number of contexts by different species, e.g., by suckling lambs or kids, by dogs when experiencing excitement and pleasure. The action is thought to function as a signal communicating the motivational state of the animal performing the behavior. Tail wagging in goats is also a sign of sexual receptivity during estrus. Occasionally it may be performed by nonestrous goats when approached by a mature male. It may also be performed by juvenile does prior to sexual maturity in response to sniffing or mounting attempts by male penmates.

TAIL WAGGING (ducks): A postural display characterized by depressed tail feathers that are rapidly moved from side to side.

TAINT, BOAR: *See* Boar Taint.

TAKE: Cooperation by a female with the sexual advances, including mounting, of a male.

TAME: Referring to an animal that has been familiarized with humans so as to be tractable.

TANESMUS: Excessive straining during defecation. Tanesmus may indicate paralysis of the anus or be a sign of rabies or meningitis.

TASTE: To perceive stimuli through the gustatory sense mode. Also, the particular gustatory characteristics of a substance.

TASTE AVERSION: Learned avoidance of a particular type of food that has caused an animal to become ill after eating it. Taste aversion exemplifies a special case of operant conditioning in which the lag between action (consumption of the food) and reinforcement (illness) is unusually long. The stimuli by which the animal is able to identify the food to be avoided are thought to involve the sense of taste.

TASTE BUD: One of many tiny protuberances on the tongue that contain taste receptors.

TATTOOING: Permanent intracutaneous marking of an organism for the purpose of individual identification.

TAXIS: Locomotion toward (positive taxis) or away from (negative taxis) some environmental factor. Such factors characteristically have an influence on homeostasis or reproduction. *Compare*: Tropism.

TEASER: In an agricultural sense, a conspecific animal used for detection of females in estrus, or an animal that serves as a mounting object during semen collection.

TEASING: A process of testing females for readiness to mate.

TEAT DIPPING: Partial submersion of the teats of lactating animals into a disinfecting fluid to prevent infection.

TEAT ORDER: A numerical system to designate the position of teats along an udder. In a looser sense, the term is used for the order of littermates along an udder during a suckling event.

TEAT SAMPLING (swine): A tendency of neonatal piglets to seek out successive teats on the dam's udder and suckle briefly from them. Teat sampling occurs during the neonursing phase on the day of birth soon after the piglet encounters its first teat and before the development of teat specificity. It is thought to enable the piglet to assess the relative quality of each teat.

TEAT SPECIFICITY: The tendency of a preweaned mammal to suckle consistently from a particular teat in the dam's udder.

TEAT SUCKING, JUVENILE: *See* Juvenile Teat Sucking.

TECHNOPATHY: Disease, injury, body deformity, or performance of behavior that is harmful to the animal or its peers, and develops as a consequence of production technology, including housing and management.

TEETH, BARED (horse): *See* Bared Teeth.

TEG: A two-year-old sheep.

TELEMETRY: Measurement, transmission, and recording of information (e.g., behavior records) by means of a remote sensor apparatus. Signals are transmitted through any of a number of modes, such as radiowaves, electrical impulses in wires, etc.

TELENCEPHALON: The anterior part of the brain, which includes olfactory lobes, cerebral cortex, and corpora striata.

TELEOLOGY: Interpretation or explanation of natural phenomena (e.g., behavior) from the viewpoint of purpose or function.

TELERECEPTOR: A sensory receptor that is adapted to respond to distal stimuli. A telereceptor does not require mechanical contact with the stimulus to be activated. *Compare*: Contact Receptor.

TELORECEPTOR: *See* Telereceptor.

TELOS: A specific quality of living organisms (in contrast to nonliving organisms) characterized by their genetically imprinted nature, activities, and functions. In ancient times the term was used by Aristotle. In contemporary animal welfare literature, B. Rollin uses the term telos to refer to intrinsic biological characteristics and corresponding needs and interests of an animal which define its own existence (i.e., the spiderness of a spider, cowness of a cow, etc.). According to this theory, denying or thwarting these biological characteristics of an animal leads to denying or thwarting its telos.

TELOTAXIS: Locomotion conducted in approximately a straight line toward or away from some environmental factor.

TEMPERAMENT: A simplified generalization of an organism according to its excitatory or inhibitory reactions, persistent habits, level of motor activity, emotionality, alertness, curiosity, etc.

TEMPERATURE: *See* specific temperature—AMBIENT; CRITICAL; EFFECTIVE ENVIRONMENTAL.

TEMPLATE, BEHAVIORAL: *See* Behavioral Template.

TEMPORAL CONDITIONING: Conditioning wherein the unconditioned stimulus or reinforcement is presented at regular intervals. It is postulated that cues provided by an internal clock elicit the conditioned response.

TENDENCY, ORAL: *See* Oral Tendency.

TENDON: A connective tissue that attaches a muscle to a bone.

TERATOLOGY: The study of malformations and related malfunctions of organisms' bodies.

TERMINAL SENSITIVITY: The highest intensity of sensation in a given sense mode that an organism is capable of experiencing. It may or may not correspond with the highest intensity of stimulation. *Compare*: Terminal Stimulus.

TERMINAL STIMULUS: The highest intensity of stimulation in a given sense mode to which the organism is capable of responding.

TERRITORIAL AGGRESSION: Aggression manifested in the context of establishment, protection, or extension of a territory.

TERRITORIAL BEHAVIOR: Behavioral actions performed by an individual, pair, or group to establish or maintain a territory (e.g., boundary marking, vocalization, visual displays, aggression, and avoidance of neighboring territories).

TERRITORY: An area that an individual, pair, or group defends and maintains free of intruders, or remains within by avoiding incursions into neighboring, occupied areas.

TEST: *See* specific test—ASSOCIATION; DRAIZE; HALOTHANE; LD-50; OPEN-FIELD.

TESTES: Male gonads.

TESTES, HIDDEN: *See* Cryptorchid.

TESTING: *See* specific testing—SIB; PROGENY.

TEST STIMULUS: An experimental stimulus. This term is most commonly used in generalization studies to determine the ability of animals to differentiate between a novel stimulus and the original stimulus.

TETANUS: A disease caused by *Clostridium tetani* and characterized by spasmodic or constant muscular rigidity. Affected animals have difficulty walking and may develop a stiff, saw-horse posture.

TETHER: The instrument or device used to tie animals to one specific location (e.g., a barn stancheon), but permitting free feed intake, standing, and lying down. Also, to tie an animal with a tether.

TETRAPLEGIA: *See* Quadriplegia.

TEXAS GATE: A grid, having intrabar spaces large enough for a hoof to pass through, placed over a ditch across a roadway at the entrance to a fenced-off area. A Texas gate will allow wheeled vehicles to pass through the entrance, but will contain horses or cattle within the enclosure because they will refuse to step on the gate.

THALAMUS: Part of the forebrain through which sensory information is transmitted to the cortex of the cerebral hemispheres.

THEORY: *See* specific theory—CONTROL SYSTEMS; DIVINE COMMAND; ETHICAL; IDEAL OBSERVER; NATURAL LAW; STIMULUS-RESPONSE; STIMULUS SUBSTITUTION; VALUE.

THEORY OF EVOLUTION: A theory that proposes that speciation has occurred as a consequence of natural selection resulting in progressive adaptation of organisms to differing environments.

THERAPY: Treatment of disease. *Also see* specific therapy—AVERSION; BEHAVIOR.

THERIOMORPHISM: Representation of supernatural entities in the form of, or incorporating characteristics of, animals. Also, religious interpretation of animals and animal characteristics as supernatural.

THERMAL: Pertaining to the sense of temperature. The term also refers generally to temperature (e.g., thermal environment, thermal measurement, etc.)

THERMAL COMFORT ZONE (TCZ): The range of environmental temperature in which an endothermic organism can maintain a stable body temperature without metabolic heat production above that produced in the course of maintenance metabolism, and without employing any behavioral strategy to conserve or dissipate heat. The actual temperature range of the thermal comfort zone depends on many factors, such as air movement, humidity, solar radiation, bedding, health, age, nutrition, and body insulation.

THERMONEUTRAL ZONE (TNZ): The range of environmental temperature in which an endothermic organism can maintain a stable body temperature without metabolic heat production above that produced in the course of maintenance metabolism. The thermoneutral zone is somewhat broader than the thermal comfort zone because the animal is able to adopt energetically inexpensive behavioral strategies to conserve or dissipate heat.

THERMORECEPTOR: A sensory receptor activated by thermal stimulation.

THERMOREGULATION: A variety of physiological and behavioral adjustments that assist an animal to maintain a stable body temperature.

THETA ACTIVITY: *See* Brain Wave Activity.

THIGMOTAXIC BEHAVIOR: Tendency to stay close to or in contact with walls or objects and to avoid open areas. This tendency is relatively common in many species of animals during shelter seeking and resting.

THINK: To engage in mental analysis, integration of information, and generation of ideas, conclusions, or decisions.

THIRST: An uncomfortable sensation induced by deprivation of liquids. Thirst generally is associated with increased appetite for liquids, provided the duration of such deprivation has not resulted in lethargy.

THOROUGHBRED: An English breed of running horses.

THREAT: Indication of intent to harm directed toward an adversary. Functionally, threat provides an opportunity for the threatened individual to resolve the situation by escape, avoidance, or display of submission.

THREAT SIGNAL: Any signal indicative of threat. Depending on the species, such signals may include ritualized fight movements, pawing with the forelegs, foot stamping, display of weapons, approach to the limit of the opponent's individual distance zone, growling, snorting, etc. *Compare*: Warning Signal.

THREE-DAY-EVENT (horse): An equestrian competition combining dressage (first day), speed and cross-country riding (second day), and show-jumping (third day).

THREE-GAITED (horse): A horse trained to execute walk, trot, and canter for pleasure riding or showing. In the American Saddlebred, three-gaited horses have sheared manes. *Compare*: Five-gaited.

THRESHOLD: The minimum or maximum stimulus intensity that an organism is able to sense. *Also see* specific threshold—DIFFERENCE; EMIGRATION; STIMULUS.

THRIFTY: Of robust health and vigor. (colloquial term)

THROW: The act of causing an animal to fall to the ground for the purpose of branding or some other treatment.

THRUSTING: Repeated, forceful movement forward followed by slower retraction. Such action may be performed with the pelvis, head, tongue, etc.

THUMPING (rabbit): Simultaneous striking of both hindpaws on the ground. Thumping is considered to be a warning signal.

THYROID STIMULATING HORMONE (TSH): A hormone of the anterior lobe of the pituitary gland that stimulates the activity of the thyroid gland in its secretion of thyroid hormones.

TIDBITTING (chicken): Calls produced by mature males and broody females consisting of a series of regularly spaced, wide frequency range sounds. They are vocalized by males during courtship and aggressive encounters, and by females in an epimeletic context.

TIE STALL: A stall within a housing unit where an animal is tethered except when released temporarily for exercise, grazing, milking, or examination.

TIMER, CIRCADIAN: *See* Circadian Timer.

TIMER, INTERVAL: *See* Interval Timer.

TIME RELEASER: *See* Time Setter.

TIME SETTER: A stimulus that instigates some time-dependent occurrence or synchronization of activities (e.g., on a daily, monthly, seasonal, or yearly basis).

TIME-SHARING: Alternation between two behavioral actions due to disinhibition and motivational competition. When the motivation for a dominant activity declines, the subdominant activity is disinhibited and is manifested until rising motivation for the dominant activity inhibits the subdominant activity through motivational competition.

TIME SYNCHRONIZER: *See* Time Setter.

TIMID ANIMAL: Omega animal. (colloquial term)

TOE TRIMMING (poultry): Removal of the distal portion of the phalanges, including the toenail.

TOLERANCE: The ability of an organism to survive in a given range of environmental conditions and maintain normal bodily functions. *Also see* specific tolerance—SOCIAL.

TOLERANCE REFLEX: A technical term referring to a response of a female, characterized by standing still, flexion of the back, and lifting of the tail to one side when the lumbo-sacral region is massaged or pushed on by an attendant. This response indicates sexual receptivity.

TOM: A mature male turkey or mature male cat.

TONGUE: A muscular, mobile structure attached at the ventral posterior of the mouth. The tongue is used for manipulation of food in the mouth during mastication, for licking, and to assist vocalization. The dorsal surface of the tongue has taste buds that contain gustatory receptors which mediate the sense of taste.

TONGUE, CURLED: *See* Curled Tongue.

TONGUE LOLLING: Protrusion of the tongue hanging from the oral cavity, not associated with eating, drinking, or searching for food. Tongue lolling may develop into a habit that, in show horses, is considered a nuisance vice.

TONGUE PLAYING: *See* Tongue Rolling.

TONGUE ROLLING: Repeated extension of the tongue followed by rolling of the tip back into the mouth. Tongue rolling occurs more frequently in cattle fed a high concentrate diet and is considered to be a nuisance vice.

TONGUING: A distinct form of scratch reflex manifested in bovines as periodical insertion of the tongue into a nostril.

TONIC IMMOBILITY: Temporary inability to move manifested by an animal upon close exposure to, or handling by, some powerful fear-producing agent (e.g., "feigning of death" in the proximity of a predator).

TONUS: Sustained low level of muscular contraction to maintain bodily posture and physiological homeostasis.

TOOL: Any object used as an extension of the body to facilitate achievement of a given purpose.

TORPOR: A state of minimized metabolic and physical activity and sensation associated with lowered body temperature. Total inactivity.

TOSSING, HEAD: *See* Head Tossing.

TOUCH: To come into contact with another organism, object, or part of the body. The term usually is used for situations in which the contact does not involve a great deal of force.

TOXIC: Poisonous.

TRACE CONDITIONING: A type of classical conditioning in which the conditioned stimulus is presented and terminated prior to the onset of the unconditioned stimulus.

TRACT, CORTICOSPINAL: *See* Corticospinal Tract.

TRACTABLE ANIMAL: An animal that is docile and can be easily controlled.

TRAIL: A pathway made by animals walking to pastures, water source, or shelters. Animals within an established group have strong tendency to follow each other and use the same route.

TRAINABILITY: Capacity of an organism to be trained to perform a given task. Trainability may refer to quantitative aspects of training (e.g., time required to reach a certain speed of response) or qualitative aspects (e.g., ability of response).

TRAINING: Any procedure instituted to cause acquisition of a required action or inhibition of undesirable action by an animal that will enable the animal to make an appropriate response in a given circumstance. The term training also is used for progressive conditioning (e.g., behavior shaping in equine dressage) or physical exercise to improve some aspect of performance. *Also see* specific training—DISCRIMINATION; DISCRIMINATIVE; DISCRIMINATIVE PUNISHMENT; DISCRIMINATIVE REWARD; ODDITY; PUNISHMENT; REVERSAL; REWARD.

TRAIT, DOMESTICATION: *See* Domestication Trait.

TRANQUILIZER: A chemical compound (drug) that reduces anxiety or fear and increases relaxation.

TRANSFER, EMBRYO: *See* Embryo Transfer.

TRANSFER PROCESS: A hypothetical process that incorporates all the effects of previous experiences on the performed behavior.

TRANSMISSION, CULTURAL: *See* Cultural Transmission.

TRANSPLANT: Any biological material (e.g., organ, tissue, cell, blood, embryo, etc.) taken from the body of one organism (donor) and artificially transferred to another organism (recipient) to become part of the recipient's body and fulfill the same or similar function as it had in the donor.

TRAUMA: Any experience of serious physical or psychological suffering that has a lasting effect on an organism.

TRAVERS (horse): A dressage maneuver of the Spanish High Riding School. The horse bends slightly around the inside leg of the rider and moves forward while looking in the direction of movement. The outside legs cross in front of the inside legs.

TREADING (chicken): Behavior of males displayed in the copulatory mounting position prior to cloacal contact, characterized by several relatively fast treading movements of the legs.

TREMBLING: Involuntary, rapid, phasic contraction of muscles.

TREMOR: *See* Trembling; Epidemic Tremor.

TRIAL: A unit of practice in learning experiments during which an organism is exposed to a specific stimulus situation.

TRIAL AND ERROR LEARNING: A type of learning whereby an organism gradually adopts responses that are most successful in a given learning situation.

TRICHROMATIC THEORY: A theory that perception of all colors is based on three primary colors.

TRIGEMINAL NERVE: The fifth cranial nerve which innervates the nasal cavity and the muscles of mastication.

TRILL (chicken): High amplitude, repeated calls (~3 per second) produced by chicks in response to unfamiliar noises, strange objects and sudden changes in environment. Trilling appears to indicate a state of alarm.

TRIUMPH CEREMONY (geese): Symbolic attack toward an aversive stimulus followed by return either to the main body of the group or to partner or prospective mate, after which the group or pair engage in intense vocalization, neck stretching, and wing-flapping.

TROCAR: A sharp pointed stylet inside a metal tube that can penetrate abdominal muscles, peritoneum, and the wall of a rumen. When the stylet is withdrawn after penetration, the tube facilitates release of gases from a bloated digestive tract.

TROCHLEAR NERVE: The fourth cranial nerve which innervates the superior oblique eye muscle.

TROPHALLAXIS: Mutual or unilateral exchange of alimentary fluid among individuals. Trophallaxis is most common in colonies of social insects.

TROPHIC: Pertaining to nutrition.

TROPISM: Intrinsic orientation toward (positive tropism) or away from (negative tropism) some environmental factors, characteristically those related to homeostasis. *Compare*: Taxis.

TROT: A naturally developed regular two-beat gait with the leg movement synchronized diagonally. The sequence of hoof beats is left hind together with right fore followed by right hind together with left fore.

TROTTER (horse): A horse that races using a two-beat diagonal gait called the trot.

TRUE ESTRUS: *See* Receptive Phase.

TUBERCULOSIS: A contagious bacterial disease of all domesticated animals, many wild animals, and humans, caused by *Mycobacterium tuberculosis*.

Behavioral symptoms of this chronic disease develop very slowly, from increased coughing (in the case of lung tuberculosis) to a totally weakened body constitution.

TUPPING (sheep): Mating and insemination in sheep. (colloquial term)

TUSHES: Canine teeth in horses.

TUSKS: Canine teeth in the pig and a number of other mammalian species.

TWINS: A pair of siblings developed during one pregnancy cycle and born close together in time. *Also see* specific twins—FRATERNAL; IDENTICAL.

TWITCH: A brief contraction of skeletal muscle.

TWITCH (horse): A simple restraint device consisting of a flexible loop (e.g., rope) connected to a short stick. The loop, when placed around a horse's lip, can be twisted gradually to squeeze the lip, thus controlling the horse's attention and promoting handling ease.

TWITTER: A relatively low-amplitude, repeated vocalization produced by chicks in response to certain novel or suddenly appearing stimuli. Generally, each element in a twitter series is characterized by short duration and rapidly ascending sound frequency. Twittering seems to indicate a state of mild arousal without apparent fear or distress.

TYING-UP (horse): *See* Exertional Myopathy.

TYMPANIC CAVITY: An air-filled space behind the tympanic membrane in the middle ear that contains the tympanic ossicles and, by balancing the air pressure across the tympanic membrane, allows the membrane to vibrate properly in response to sound waves.

TYMPANIC MEMBRANE: A thin membrane at the base of the meatus that separates the external ear from the middle ear. The tympanic membrane vibrates when sound waves impinge on it and transfers this vibration to the tympanic ossicles. Synonym: Ear Drum.

TYMPANIC OSSICLES: Three small bones of the inner ear (stirrup, stapes, anvil) that transmit and magnify vibrations from the tympanic membrane to the inner ear.

TYMPANY, RUMINAL: *See* Bloat.

TYPE: The conformation of an animal's body. For the purpose of evaluation, animals are compared to the ideal type for a given breed or strain.

TYPING, BLOOD: *See* Blood Typing.

U

UDDER: The mammary gland of females in some mammalian species.

ULCER: Any long lasting breach in the skin or in the membrane of the body cavity.

UMWELT: Surroundings as perceived by an organism.

UNCERTAINTY INTERVAL: The range of intensities of stimulation perceived as one level of sensation by an organism.

UNCONDITIONAL STIMULUS: *See* Unconditioned Stimulus.

UNCONDITIONED REINFORCER: *See* Primary Reinforcer.

UNCONDITIONED RESPONSE: Any unlearned response that is directly elicited by an unconditioned stimulus (e.g., salivation in response to presentation of food).

UNCONDITIONED STIMULUS: Any stimulus that invariably elicits, without association with any other stimulus, a specific response from an organism, provided the motivational state of the organism is appropriate.

UNCONSCIOUS BEHAVIOR: Any behavioral action wherein the organism is aware of neither its performance nor its causation or function at the time of performing the action. This term is often inappropriately used for behaviors guided by subconscious (extraconscious) processes.

UNCONSCIOUS PROCESS: A process that is inaccessible to the conscious mind.

UNDERSTIMULATION: *See* Hypostimulation.

UNDOMESTICATED ANIMAL: Any animal whose progenitors have not undergone a domestication process.

UNGULATES: Hoofed animals.

UNIT, SOCIAL: *See* Social Unit.

UNIVOCAL: A statement or occurrence that has only one meaning. In a behavioral context this term is used in motivational assessment to refer to any behavioral action that has a singular explanation and distinct function.

UNNECESSARY SUFFERING: Any suffering that is not essential for a purpose of sufficient importance, or that can be avoided by adopting alternatives that would achieve the same purpose.

UNTAME: Referring to an animal that has not been familiarized with humans and thus is intractable, producing avoidance or aggressive responses in the presence of people.

UPPER CRITICAL TEMPERATURE: *See* Critical Temperature.

UPSIDE-DOWN POSTERIOR PRESENTATION: Fetal presentation in which the hind legs enter the birth canal while the dorsal part of the body of the fetus is oriented ventrally.

UPSIDE-DOWN PRESENTATION: Fetal presentation in which the forelegs and the nose enter the birth canal while the dorsal part of the body of the fetus is oriented ventrally.

URETHRITIS: Inflammation of the urethra, characterized by symptoms indicative of pain when the animal urinates (e.g., back arching, stiffness, interrupted urination, etc.).

URINATION: Discharge of urine from the body. Animals, particularly male horses, spread apart their hindlegs and remain stationary during urination. Bovine males and boars may occasionally urinate while walking. Swine tend to urinate in wet areas of the pen. Synonym: Micturition.

URINE SAMPLING: Nasal contact by males with the urine of urinating females, usually followed by lip curling. Urine sampling is most common in bovines.

UROGENITAL: Pertaining to the urinary and reproductory tracts.

UROLITHIASIS: A disease occurring most frequently in heavily grain-fed rams, caused by formation of stones in the urinary system, and characterized by straining, arching of the back, and grating of the teeth. *Compare*: Urethritis.

UROPYGIAL GLAND: The oil gland located above the tail vertebrae of birds.

UTERINE LIFE: The period of life during which an offspring develops in the uterus of its dam.

UTERINE MILK: Histotroph. (colloquial term)

UTERINE PROLAPSE: Partial eversion of the uterus from the vulva. Uterine prolapse occurs frequently in hens, more often in cattle than in horses, and very rarely in pigs.

UTERUS: The portion of the reproductive tract in female mammals where the fertilized ovum implants and develops through the embryo and fetus stages until parturition.

UTILITARIANISM: A theory of moral philosophy advocating the maximization of positive consequences of human actions (e.g., pleasure and happiness or satisfaction of interests and preferences) and minimization of negative consequences (e.g., suffering, deprivation) when summed over all relevant individuals. In the context of animal welfare, utilitarianism has been proposed as a guide for treatment of animals by taking into consideration the interests of both humans and animals.

V

VACUUM ACTIVITY: A behavior, typically a modal action pattern, manifested in the absence of the stimulus which normally activates such behavior. For example pregnant sows close to parturition engage in movements characteristic of nest building even if housed in farrowing crates without any nest material available.

VACUUM, OCCUPATIONAL: *See* Occupational Vacuum

VAGUS NERVE: The tenth cranial nerve which innervates the external ear, larynx, pharynx, heart, lungs, stomach, and some other parts of the body.

VALUE: *See* specific value—INSTRUMENTAL; INTRINSIC; INHERENT.

VALUE THEORY: A philosophical theory that attempts to provide a basis for judging whether some entities or circumstances are good or bad. Some value theories propose that just as objects have a certain mass or shape, they also have a value. Other theories define values in relation to the consequences of having positive or negative feelings that objects or situations engender. The study of value theory is called axiology.

VALVE, AGGRESSIVITY: *See* Aggressivity Valve.

VANDENBERG EFFECT: Enhancement of sexual maturation in a female by exposure to male pheromones.

VARIABLE APPETITE: A clinical term for irregular appetite caused by the progression of a disease.

VARIABLE INTERVAL REINFORCEMENT SCHEDULE: A reinforcement schedule in which the reinforcing stimulus is never presented before a a variable length of time after the previous reinforcing stimulus. The length of time varies within a specified range. Antonym: Fixed Interval Reinforcement Schedule.

VARIABLE RATIO REINFORCEMENT SCHEDULE: A reinforcement schedule in which the occurrence of the reinforcing stimulus is contingent upon completion of a variable number of responses. The number of required responses for each reinforcement varies within a specified range. A special form of this schedule is called the progressive ratio reinforcement schedule, in which a progressive increase in the number of responses is required for each consecutive reinforcement. Antonym: Fixed Ratio Reinforcement Schedule.

VASCULAR: Pertaining to blood vessels.

VASECTOMY: Surgical severing of the vas deferens to cause infertility without affecting sexual desire.

VASOPRESSIN: *See* Antidiuretic hormone.

VEAL CRATE: *See* Crate.

VEALER: A calf raised for slaughtering at an early age.

VEGETATIVE ACTION: A term occasionally used to refer to nonreproductive biological processes such as respiration, eating, or excretion.

VEGETATIVE NERVOUS SYSTEM: The autonomic nervous system. (The term is no longer widely used.)

VENTRAL: At or in the direction of the lower part of the abdomen.

VENTRAL RECUMBENCY: Lying during which the sternal part of the thorax and ventral part of the abdomen are in contact with the ground. The legs are typically folded close to the body or extended cranially. Ventral recumbency often is displayed in cold ambient temperature, particularly in young animals, because it reduces the exposed body surface area.

VENTRAL TRANSVERSE PRESENTATION: Fetal presentation in which the body of the fetus is oriented in a transverse position while all four limbs and the nose enter the birth canal.

VENTRIFLEXION: Bending movement induced by flexion of ventral muscles and relaxation of dorsal muscles, (e.g., lowering of an animal's head). *Compare*: Dorsiflexion.

VESTIBULAR STIMULATION: Stimulation received by neonatal and young organisms as a result of handling and other movements instigated by parents, primarily the mother (or surrogates), and siblings. Such stimulation elicits a variety of responses, e.g., orienting, righting, following, etc., and apparently facilitates proper psychomotor development of the young organism. Prevention of vestibular stimulation may lead to abnormal ontology of motor behavior and result in kinetic disturbances (e.g., stereotypy).

VESTIBULE: A structure of the inner ear that connects with both the tympanic ossicles and the cochlea and mediates transmission of sound vibrations from the former to the latter.

VIABILITY: Ability to survive.

VICARIOUS LEARNING: *See* Observational Learning.

VICE: An undesirable habit of an animal. Dangerous vices can be harmful to the animal itself, to peers, or to humans. Nuisance vices are esthetically unpleasing to humans.

VIGILANCE: The characteristic or condition of being attentive to threat and danger. Behavioral signs of vigilance are raised head and sensory focusing to survey the surroundings.

VIGILANT BEHAVIOR: Any behavior indicative of alertness to potential danger. *Compare*: Sentinel Behavior.

VIRILE SYNDROME: Progressive masculinization occurring in females.

VIROSIS: Any virus-caused disease.

VISCEROCEPTOR: A sensory receptor activated by stimulation arising in the viscera.

V

VISION: Perception of light stimuli. Vision is mediated by the specialized structures and visual receptors in the eye. Synonym: Sight.

VISUAL: Pertaining to the sense of vision.

VISUAL ACUITY: Capacity for distinguishing visual detail.

VISUAL CLIFF: Any situation in which an animal perceives a sudden drop-off in the substrate, even if the support under foot continues to be horizontal and solid. Such situations are used in the study of perception. Animals of some species or age will refuse to cross over onto a solid but transparent floor beyond a visual cliff.

VISUAL FIELD: The complete area, including all objects in it, visible to the eye at any given time.

VISUAL ILLUSION: *See* Optical Illusion.

VIVIPAROUS: Referring to organisms having the embryo and fetal stages develop inside the mother and the offspring delivered by parturition. *Compare*: Oviparous.

VIVISECTION: Cutting into a living animal. This term commonly refers to any surgery on living animals practiced for the purpose of research.

VOCALIZATION: Production of sounds by the vibration of vocal cords in the larynx. The sounds may be modified by the organs and structures of the oral cavity. *Also see* specific vocalization—"MM"; "MOO"; "MOOEH"; "MOOH"; "MRRH."

VOLITION: Cognitive processes involved in decision making and pursuit of objectives.

VOLTE (horse): A dressage maneuver in which the horse follows a circle 6 m in diameter.

VOLUNTARY BEHAVIOR: Any action mediated by volition.

VOMIT: To eject gastrointestinal contents through the mouth. Vomiting is controlled by the vomiting center of the medulla and is elicited by simultaneous contraction of the diaphragm and abdominal muscles, and relaxation of the esophageal sphincter.

WAGGING, TAIL: *See* Tail Wagging

WAGGLE DANCE (bees): A repetitive behavior pattern performed on a vertical surface consisting of movement forward in a short straight line while moving the abdomen from side to side followed by movement alternatively to the left or right in a semicircle to return to the starting point of the straight line. The angle of the straight line to a vertical line corresponds to the angle between the position of the sun and the food source.

WALK (horse): A naturally developed, evenly cadenced, four-beat gait. The sequence of hoof beats are left hind, left fore, right hind and right fore. Depending on the speed, two or three hooves may be touching the ground at any one time.

WALKING: Relatively low speed locomotion of an organism on the ground in which propulsive force derives from the action of legs. *Also see* specific walking—STALL.

WALLOWING: An activity (considered to be comfort related) characterized by partial submersion of the body into some wet substrate (e.g., puddle, mud, etc.). Among farm animals, wallowing is most common in swine and is positively correlated with environmental temperature.

WALTZING (chicken): A postural display of males performed as a semicircular or circular locomotion around hens or male opponents. The outside wing is usually lowered, sometimes so that its primaries touch the ground and produce a rasping sound. Waltzing can be a part of a threat or courtship behavior.

WANT: *See* Desire.

WARBLES: A parasite disease of cattle, horses, and some other mammalian species caused by the hypoderma fly *(Hypoderma bovis, H. lineata)*. Larvae from eggs inserted into the skin of the affected animal migrate first to the gullet and later to the dorsum where they form marble size swellings, each containing a maggot. Occasionally the parasite causes injury to the spinal cord. Animals display fear and attempt to escape from the buzzing sound of the flying insect. Affected animals may show disorders caused by spinal cord damage, and lactating females have reduced milk yield.

WARNING CALL: A vocal warning signal.

WARNING SIGNAL: Any signal broadcast by an organism to warn others of its presence and of its intent to be aggressive in the event of an incursion (e.g., territorial calling in the context of territorial maintenance). The term also includes threat signals directed toward an opponent(s) during a situation of social conflict. *Compare*: Threat Signal.

WATER BAG: The appearance of fetal membranes prior to parturition. (colloquial term)

WATERING: Delivery of drinking water to animals.

WATER SHYNESS: Excessive or persistent refusal to enter into water. This behavior is common in animals which have not had prior opportunity to investigate bodies of water.

WAXY-SEAL (horse): Preparturient occurrence of waxy substances on the tips of the teats. Indicates impending parturition. (colloquial term)

WEANING: Permanent discontinuation of nursing of young mammals.

WEANLING: Any young animal that recently has been weaned.

WEAPON: Any organ of the body able to injure an adversary or prey.

WEAVING (horse): Stereotyped transfer of body weight from one side to the other in conjunction with waving of the head back and forth.

WEAVING (swine): Stereotyped, rhythmical swaying of the head. It occurs most often in individually housed pigs on concrete floors without bedding material when the animals are fed high concentrate diets. It is thought to be an indicator of boredom.

WELFARE: *See* Well-being.

WELL-BEING (animal): A state of harmony between the animal and its environment, characterized by optimal physical and psychological functioning and high quality of animal's life. An animal's environmental requirements for well-being may change over time (e.g., due to age, reproduction cycle, experience). The most commonly used indicators of well-being are absence of symptoms indicative of hypostimulation or hyperstimulation, physical and psychological health, and longevity. *Compare*: Quality of Life.

WETHER: A male sheep castrated before sexual maturity.

WHETTING: A colloquial term occasionally used to refer to repeated stroking or rubbing actions of some parts of animal body (e.g., horns, teeth) against surrounding objects.

WHINING (chicken): A type of call characterized by low sound frequency, pulse-like vocalization with an extended final tone in each sequence. It is assumed that whining calls have threat and sexual functions.

WHINNY (horse): A medium amplitude, rapidly pulsating sound of intermediate duration (0.5-1.5 sec) and descending pitch. A horse begins the vocalization with its mouth closed but ends with it open. It is produced by horses that are separated from their peers and also may be a greeting signal, as it often is emitted upon sight of a peer or a caretaker. The whinny is thought to be a sign of mild excitement.

WHISTLING: A sound during respiration caused by inflammation or thickening of mucus membranes lining the pharynx.

WHITE SCOUR: *See* Colibacillosis.

WHITTEN EFFECT: A phenomenon occurring when presence of a male has a positive, stimulatory effect on manifestation of estrus in conspecific females.

WHITTEN PHENOMENON: *See* Whitten Effect.

WILD ANIMAL: An undomesticated animal, an animal not living under human control, or an untame animal.

WIND SUCKING: *See* Cribbing.

WING-FANNING (bees): A rapid oscillating movement of the wings by a grounded bee in front of the entrance to a beehive. It is assumed that wing-fanning has a primarily communicative purpose, helping foraging worker bees to identify their own hives, and aids in thermoregulation of the hive environment.

WING-FLAPPING (chicken): A postural display of birds performed as rapid, and often loud, synchronized movement of wings. Wing-flapping occurs on hot humid days and most frequently immediately after mating.

WINGING-OUT (horse): Excessive paddling.

WING SHOULDER: A postural adjustment to minimize weight pressure on lateral digits of the feet. It may be a symptom of laminitis or hypophosphataemia.

WING STRETCHING (chicken): Simultaneous extension of the wing and leg on the same side of the body, assumed to be a comfort movement.

WINKING (horse): Rhythmical contraction and relaxation of vulval labia and eversion of the clitoris. Winking occurs most frequently during estrus and, normally, for a short time after urination.

WISDOM, NUTRITIONAL: *See* Nutritional Wisdom.

WITCH'S MILK: Milk secretion that occurs occasionally in small amounts by neonates. (colloquial term)

WITH CALF (cattle): *See* In Calf.

WITHER NIBBLING (horse): Reciprocal allogrooming jointly performed between horses facing in opposite directions. Both partners gently bite and lip stroke each other's withers and other dorsal areas of their bodies.

WITHHELD RESPONSE: In a binary classification relevant to operant conditioning, a response manifested as the nonperformance of a particular motor action. For example if a horse is being trained to jump, the nonmanifestation of the motor action of jumping when the cue to jump is given is considered to be a withheld response. Antonym: Produced Response.

WOBBLER (horse): A young horse that shows swaying action of the hindquarters, stumbling, and distorted movements, particularly when trotting. Spinal injury is the main cause of these locomotory defects.

WOOD GNAWING: Gnawing of wooden fences, partitions, or other fixtures by horses, cattle, pigs, and rabbits. Wood gnawing is a fairly common behavior of farm animals but its occurrence is increased when animals are fed high concentrate diets.

WOOL PICKING: *See* Fleece Pulling.

WOOL PULLING: *See* Fleece Pulling.

WORK (horse): Activity of a horse directed by humans (e.g., pulling equipment, rounding up cattle, etc.).

WOUND: Any induced breach in the continuity of a tissue.

WRY-NECK: Deformity in a horse or cattle fetus characterized by rigidly curved neck that resists intrauterine effort to straighten it and generally prevents passage of the fetus through the birth canal.

Y

YAWN: An involuntary gaping of the mouth accompanied by inspiration of air and, often, stretching of the body.

YEANING: The process of giving birth in sheep and goats. Synonym: Kidding, Lambing.

YEANLING: A young sheep or goat.

YEARLING: An animal in its second year of life.

YELD MARE (horse): A mare that fails to give birth to an offspring in a given year.

YELLOW BODY: *See* Corpus Luteum.

YERKES-DODSON LAW: A law that says that the motivation for learning a task decreases with increased complexity of the task.

YIELD: Production performance of cattle (milk yield), poultry (egg yield), etc., over a given period of time.

YOKED CONTROL: The individual in a yoked experiment that receives reinforcement contingent on a given operant response of another individual(s), while its own performance of the same response does not result in its reinforcement.

YOKED EXPERIMENT: An experiment in which the technique of yoked reinforcement is used, generally to separate the effect of reinforcement frequency from the effects of other variables.

YOKED REINFORCEMENT: Reinforcement of one or more organisms that is contingent on a given operant response by another organism. For example, if an animal located in one compartment performs a given response, an animal in a second compartment is presented with a reinforcement, regardless of its own responses.

Z

ZEITGEBER: *See* Time Setter.
ZONE: *See* specific zone—COMFORT; GROUP DISTANCE; INDIVIDUAL DISTANCE; THERMAL COMFORT; THERMONEUTRAL.
ZOOMORPHISM: The attribution of the characteristics, abilities, or behavior of nonhuman organisms to humans.
ZOOPHOBIA: An excessive fear of animals by humans.
ZOOSEMIOTICS: The study of animal communication.
ZUGUNRUHE: *See* Migration Excitement.
ZYGOTE: A reproductive cell formed by the union of male and female gametes.